AF396461

Series Editor

Steven G. Krantz, Department of Mathematics, Washington University,
Saint Louis, USA

This series includes titles in applied mathematics and statistics for cross-disciplinary STEM professionals, educators, researchers, and students. The series focuses on new and traditional techniques to develop mathematical knowledge and skills, an understanding of core mathematical reasoning, and the ability to utilize data in specific applications.

Sujaul Chowdhury · Md. Golam Moktadir

Numerical Solutions Using the Taylor Series Method

Initial and Boundary Value Problems

 Springer

Sujaul Chowdhury
Department of Physics
Shahjalal University of Sci. and Tech.
Sylhet, Bangladesh

Md. Golam Moktadir
Department of Physics
Shahjalal University of Sci. and Tech.
Sylhet, Bangladesh

ISSN 1938-1743   ISSN 1938-1751   (electronic)
Synthesis Lectures on Mathematics & Statistics
ISBN 978-3-032-26991-1   ISBN 978-3-032-26992-8   (eBook)
https://doi.org/10.1007/978-3-032-26992-8

This Springer imprint is published by the registered company Springer Nature Switzerland AG
The registered company address is: Gewerbestrasse 11, 6330 Cham, Switzerland

If disposing of this product, please recycle the paper.

Preface

This book deals with the Taylor series method for the numerical solution of initial and boundary value problems. There are two chapters: one on initial value problems and another on boundary value problems using the shooting method. We have performed symbolic computation in Mathematica®, through which we have obtained the numerical solutions.

As for initial value problems, a number of differential equations related to a number of problems in physics have been solved numerically. The equations are first-, second-, and third-order differential equations. The problems are:

- Radioactive decay
- Simple harmonic motion
- Damped harmonic motion
- Driven damped harmonic motion
- Motion of oscillators in phase space
- Cyclotron motion
- Differential equation for hyperbolic function cosh

As for oscillatory motion, we have obtained both velocity and displacement of the oscillating particle as functions of time. We found that the Taylor series method works well even with third-order differential equation. As for cyclotron motion, we were able to simulate the trajectory of an electron in a magnetic field in real space.

Regarding boundary value problems, we have reproduced the Hermite polynomials H_3, H_4, and H_5 by numerically solving two-point boundary value problems using the so-called shooting method. Use of the shooting method requires treating a boundary value problem as an initial value problem, where we have made good use of Taylor series method.

A key feature of the results presented in this book is that we can use an appreciably large increment of the independent variable and get excellent agreement with known exact analytic solution even for higher (second and third) order differential equations.

Parts of the book can be used as one of course books on computational physics or computational mathematics for final-year undergraduates in physics or mathematics. The programs in Mathematica® will be clear to course instructors, if not to students.

Sylhet, Bangladesh   Sujaul Chowdhury
Sylhet, Bangladesh   Md. Golam Moktadir
November 2025

Contents

Numerical Solutions Using Taylor Series Method: Initial Value Problems

1

1.1 Introduction to the Method

Taylor series is well known as

$$y(t+h) = y(t) + \frac{h}{1!}\frac{dy}{dt} + \frac{h^2}{2!}\frac{d^2y}{dt^2} + \frac{h^3}{3!}\frac{d^3y}{dt^3} + \frac{h^4}{4!}\frac{d^4y}{dt^4} + \frac{h^5}{5!}\frac{d^5y}{dt^5} \tag{1.1}$$

where we have truncated the series after fifth derivative term, i.e. we have retained terms containing up to the fifth derivative. If we have a first order differential equation

$$\frac{dy}{dt} = F(t,y) \tag{1.2}$$

we can obtain expressions for the first five derivatives of y, i.e. we can obtain expressions for $\dfrac{dy}{dt}$, $\dfrac{d^2y}{dt^2}$, $\dfrac{d^3y}{dt^3}$, $\dfrac{d^4y}{dt^4}$ and $\dfrac{d^5y}{dt^5}$. We can use these expressions in Eq. (1.1). As such, we get a relation between $y(t+h)$ and $y(t)$. The key point is that if value of y is known at t, the series gives us value of y at $t+h$. This point makes the series useful for solving initial value problems. Denoting $\dfrac{dy}{dt}$ by $yd1$, $\dfrac{d^2y}{dt^2}$ by $yd2$, $\dfrac{d^3y}{dt^3}$ by $yd3$, $\dfrac{d^4y}{dt^4}$ by $yd4$ and $\dfrac{d^5y}{dt^5}$ by $yd5$, iteration for numerical solution of the initial value problem is

$$y_{\text{new}} = y_{\text{old}} + hyd1 + (1/2)h^2yd2 + (1/6)h^3yd3 + (1/24)h^4yd4 + (1/120)h^5yd5 \tag{1.3}$$

© The Author(s), under exclusive license to Springer Nature Switzerland AG 2026
S. Chowdhury, M. G. Moktadir, *Numerical Solutions Using the Taylor Series Method*, Synthesis Lectures on Mathematics & Statistics,
https://doi.org/10.1007/978-3-032-26992-8_1

1.2   Numerical Solution of Radioactive Decay Law Using Taylor Series Method

Radioactive decay is governed by the differential equation

$$\frac{dy}{dt} = -\lambda\, y \tag{1.4}$$

Here y is number of un-decayed nuclei at time t and λ is decay constant. Here we have numerically solved this differential equation by Taylor series method and have compared the numerical results with known exact analytical solution

$$y = y_0\, e^{-\lambda t} \tag{1.5}$$

where y_0 is initial number of un-decayed nuclei i.e. at time $t = 0$ and we take the decay constant $\lambda = 0.01$.

We now proceed to numerically solve Eq. (1.4) with $\lambda = 0.01$. Iteration for numerical solution is given by

$$y_{\text{new}} = y_{\text{old}} + h\, yd1 + (1/2)h^2\, yd2 + (1/6)h^3\, yd3 + (1/24)h^4\, yd4 + (1/120)h^5\, yd5 \tag{1.6}$$

where we have denoted $\dfrac{dy}{dt}$ by $yd1$ given by $-0.01y$, $\dfrac{d^2 y}{dt^2}$ by $yd2$ given by $(0.01^2)y$, $\dfrac{d^3 y}{dt^3}$ by $yd3$ given by $-(0.01^3)y$, $\dfrac{d^4 y}{dt^4}$ by $yd4$ given by $(0.01^4)y$ and $\dfrac{d^5 y}{dt^5}$ by $yd5$ given by $-(0.01^5)y$. The program becomes as in program number 1.1. This is symbolic computation and is evident. Numerical solution is as in Table 1.1 and Fig. 1.1.

Figure 1.1 shows numerical solution of radioactive decay law obtained using Taylor series method using increment $h = 50$ in (a) and $h = 100$ in (b) using program number 1.1. The plot (**a**) shows data of Table 1.1. Agreements between numerical data and known exact result shown by the curves are almost exact. As is evident, we could use large increment of the independent variable (time) and get perfect agreement between numerical data and known exact result

Table 1.1 Numerical solution of radioactive decay law obtained using Taylor series method using increment $h = 50$ using program number 1.1

i	t	y (approx.)	y (exact)
1	50	606.510	606.531
2	100	367.855	367.879
3	150	223.108	223.130
4	200	135.317	135.335
5	250	82.0713	82.0850
6	300	49.7771	49.7871

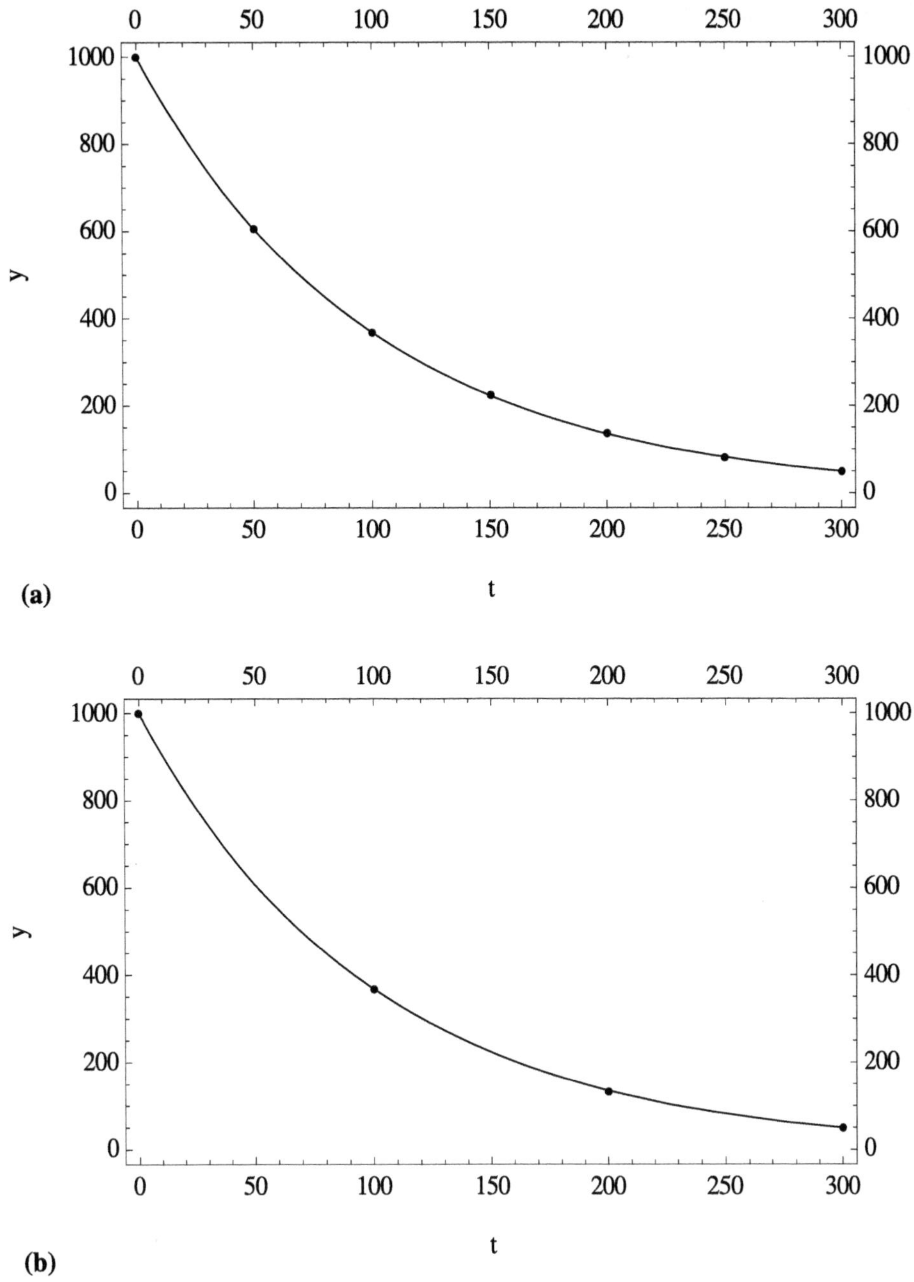

Fig. 1.1 Numerical solution of radioactive decay law obtained using Taylor series method using increment $h = 50$ in (**a**) and $h = 100$ in (**b**). The curves show known exact results

Program Number 1.1 (radioactive decay)

```
h=50;
Y[0]=y=1000;
i=0;
Table [{

i=i+1,
t=t+h,

yd1=-0.01*y;
yd2=(0.01^2)*y;
yd3=-(0.01^3)*y;
yd4=(0.01^4)*y;
yd5=-(0.01^5)*y;

Y[i]=y=y+h*yd1+(1/2)*(h^2)*yd2+(1/6)*(h^3)*yd3+
(1/24)*(h^4)*yd4+(1/120)*(h^5)*yd5,

1000*Exp[-0.01*t]},{t,0,300-h,h}];

TableForm[%,TableSpacing->{2,2},
TableHeadings->{None,{"i","t","y(approx.)","y(exact)"}}]

i=-1;
p1=ListPlot[Table[{i=i+1;t=t+h,Y[i]},{t,0-h,300-h,h}],
Frame->True,FrameLabel->{"t","y"},
FrameTicks->All,PlotStyle->{Black}];

p2=Plot[{1000*Exp[-0.01*t]},{t,0,300},PlotStyle->{Black}];
Show[p1,p2]
```

1.3 Numerical Solution of Differential Equation for Simple Harmonic Motion Using Taylor Series Method

While executing simple harmonic motion, a particle is acted on by Hooke's law force $F = -kx$ where x is displacement from equilibrium. According to Newtonian mechanics, differential equation obeyed is $m\dfrac{d^2x}{dt^2} = -kx$ where m is mass of the particle, k is spring constant and t is time. For simplicity, let m and k be equal in numerical value. As such, the differential equation becomes

$$\frac{d^2x}{dt^2} = -x \tag{1.7}$$

Here we have numerically solved differential Eq. (1.7) as an initial value problem by Taylor series method and compared the results with known exact analytical solution.

But Eq. (1.7) is a second order differential equation whereas we need first order differential equation to deal with the Taylor series. Therefore we rewrite Eq. (1.7) as two first order differential equations as

$$\frac{dv}{dt} = -x \tag{1.8}$$

and

$$\frac{dx}{dt} = v \tag{1.9}$$

To solve Eq. (1.8) using Taylor series method, iteration for numerical solution is given by

$$v_{new} = v_{old} + h\,vd1 + (1/2)h^2 vd2 + (1/6)h^3 vd3 + (1/24)h^4 vd4 + (1/120)h^5 vd5 \tag{1.10}$$

where we have denoted $\frac{dv}{dt}$ by $vd1$ given by $-x$, $\frac{d^2 v}{dt^2}$ by $vd2$ given by $-v$, $\frac{d^3 v}{dt^3}$ by $vd3$ given by x, $\frac{d^4 v}{dt^4}$ by $vd4$ given by v and $\frac{d^5 v}{dt^5}$ by $vd5$ given by $-x$.

To solve Eq. (1.9) using Taylor series method, iteration for numerical solution is given by

$$x_{new} = x_{old} + h\,xd1 + (1/2)h^2 xd2 + (1/6)h^3 xd3 + (1/24)h^4 xd4 + (1/120)h^5 xd5 \tag{1.11}$$

where we have denoted $\frac{dx}{dt}$ by $xd1$ given by v, $\frac{d^2 x}{dt^2}$ by $xd2$ given by $vd1$, $\frac{d^3 x}{dt^3}$ by $xd3$ given by $vd2$, $\frac{d^4 x}{dt^4}$ by $xd4$ given by $vd3$ and $\frac{d^5 x}{dt^5}$ by $xd5$ given by $vd4$.

We shall explore the problem in the time interval $0 \leq t \leq 6.3$. Initial values for the problem are taken as $(t, v) = (0, 0)$ and $(t, x) = (0, -1)$ implying that we have started stop watch when velocity of the particle is zero and the particle is at the extreme left in its excursion from equilibrium. The program becomes as in program number 1.2. This is symbolic computation and is evident. Resulting numerical solution is in Table 1.2 and in Figs. 1.2, 1.3, 1.4, 1.5, 1.6, 1.7, 1.8, and 1.9.

Program Number 1.2 (simple harmonic motion)

```
h=0.25;
X[0]=x=-1;
V[0]=v=0;
i=0;
```

Table 1.2 Numerical solution of differential equation obeyed by simple harmonic motion obtained using Taylor series method using increment $h = 0.25$ using program number 1.2. For initial values $(t, v) = (0, 0)$ and $(t, x) = (0, -1)$

i	t	v	x
1	0.25	0.247	−0.969
2	0.50	0.479	−0.878
3	0.75	0.682	−0.732
4	1.00	0.841	−0.540
5	1.25	0.949	−0.315
6	1.50	0.997	−0.071
7	1.75	0.984	0.178
8	2.00	0.909	0.416
9	2.25	0.778	0.628
10	2.5	0.598	0.801
11	2.75	0.382	0.924
12	3.00	0.141	0.990
13	3.25	−0.108	0.994
14	3.50	−0.351	0.936
15	3.75	−0.572	0.821
16	4.00	−0.757	0.654
17	4.25	−0.895	0.446
18	4.50	−0.978	0.211
19	4.75	−0.999	−0.038
20	5.00	−0.959	−0.284
21	5.25	−0.859	−0.512
22	5.5	−0.706	−0.709
23	5.75	−0.508	−0.861
24	6.00	−0.279	−0.960
25	6.25	−0.033	−0.999

```
Table[{
i=i+1,
t=t+h,

vd1=-x;
vd2=-v;
vd3=x;
vd4=v;
vd5=-x;

xd1=v;
xd2=vd1;
xd3=vd2;
xd4=vd3;
xd5=vd4;
```

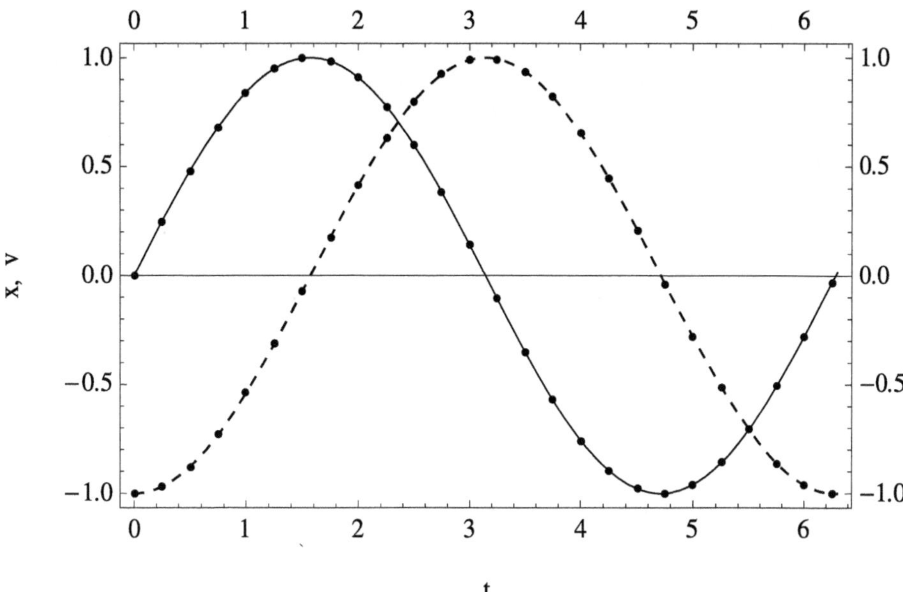

Fig. 1.2 Numerical solution of differential equation obeyed by simple harmonic motion obtained using Taylor series method using increment $h = 0.25$. Known exact results are dashed curve for displacement and un-dashed curve for velocity of the particle

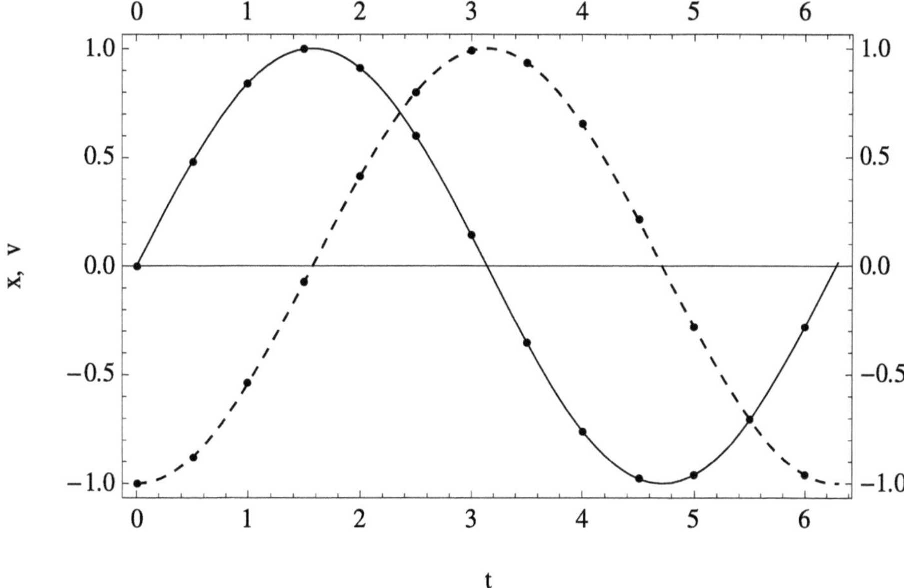

Fig. 1.3 Numerical solution of differential equation obeyed by simple harmonic motion obtained using Taylor series method using increment $h = 0.5$. Known exact results are dashed curve for displacement and un-dashed curve for velocity of the particle

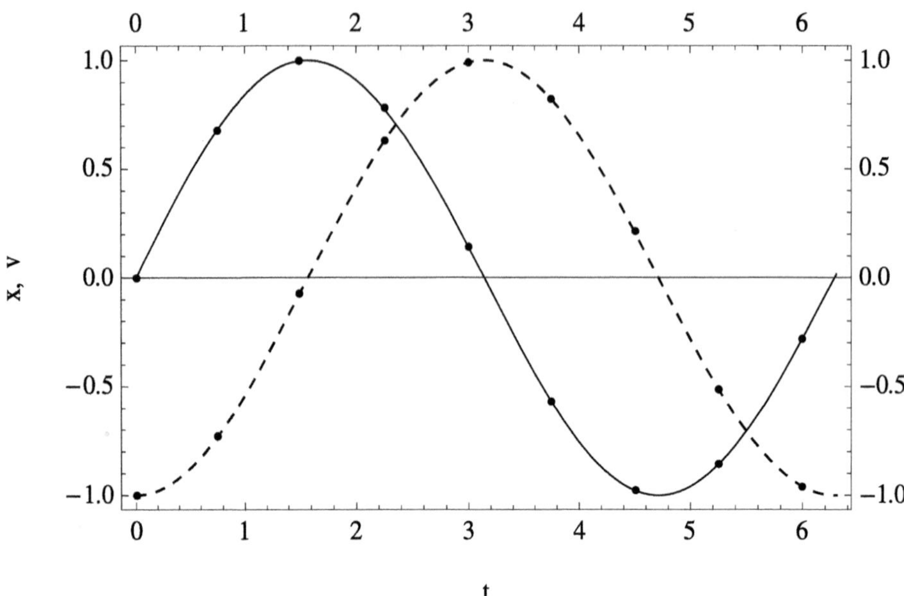

Fig. 1.4 Numerical solution of differential equation obeyed by simple harmonic motion obtained using Taylor series method using increment $h = 0.75$. Known exact results are dashed curve for displacement and un-dashed curve for velocity of the particle

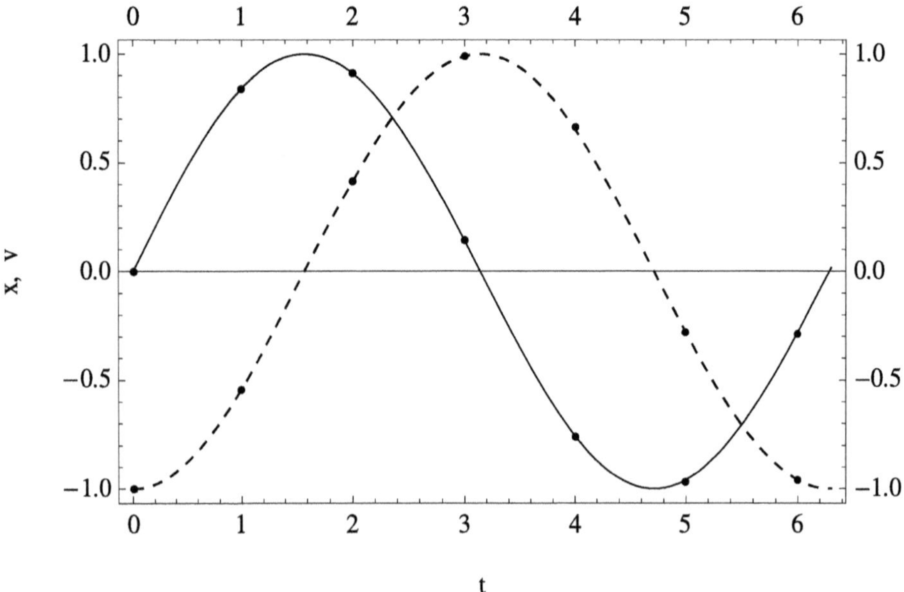

Fig. 1.5 Numerical solution of differential equation obeyed by simple harmonic motion obtained using Taylor series method using increment $h = 1.00$. Known exact results are dashed curve for displacement and un-dashed curve for velocity of the particle

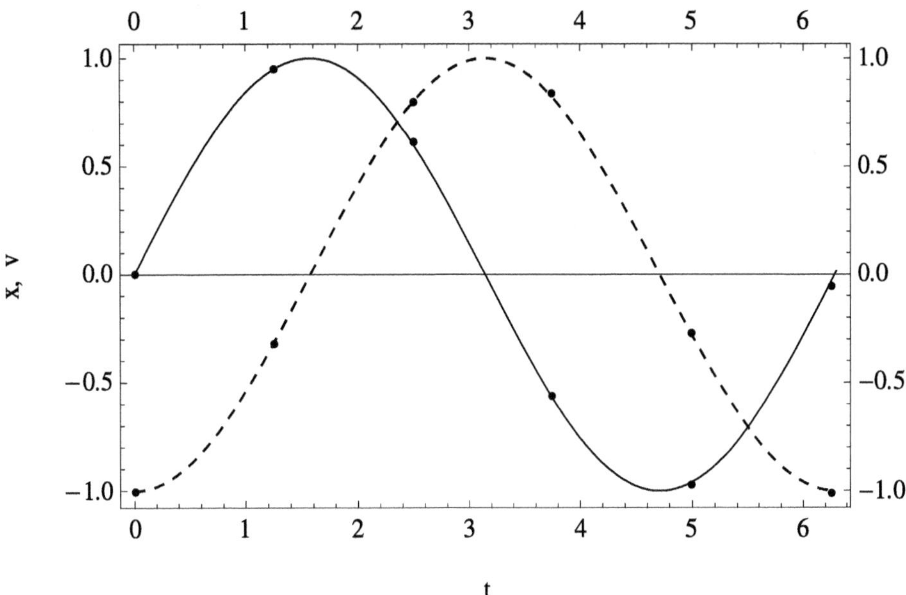

Fig. 1.6 Numerical solution of differential equation obeyed by simple harmonic motion obtained using Taylor series method using increment $h = 1.25$. Known exact results are dashed curve for displacement and un-dashed curve for velocity of the particle

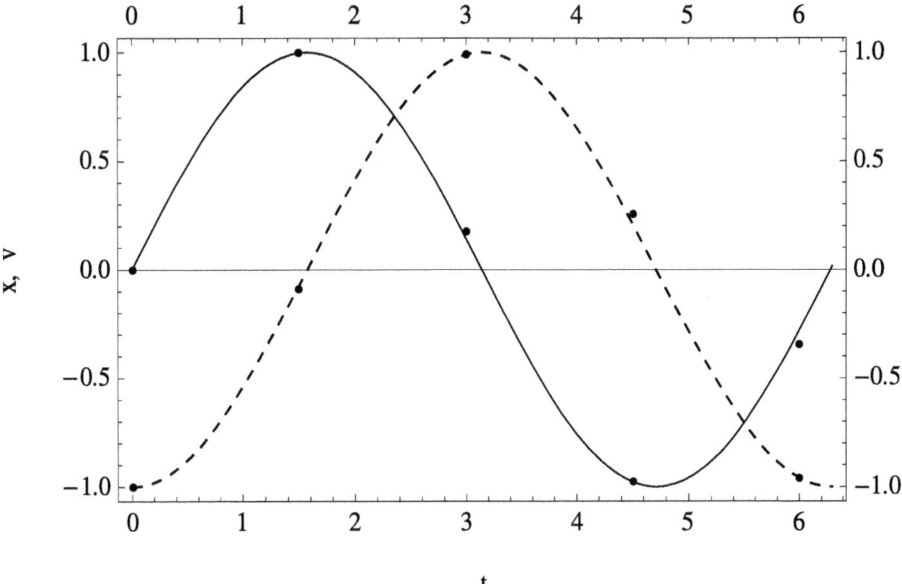

Fig. 1.7 Numerical solution of differential equation obeyed by simple harmonic motion obtained using Taylor series method using increment $h = 1.50$. Known exact results are dashed curve for displacement and un-dashed curve for velocity of the particle

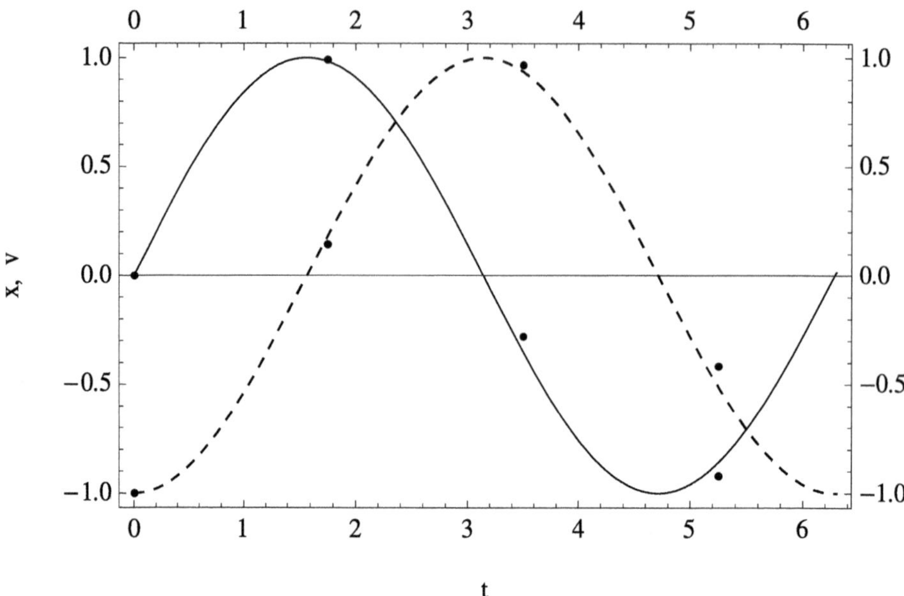

Fig. 1.8 Numerical solution of differential equation obeyed by simple harmonic motion obtained using Taylor series method using increment $h = 1.75$. Known exact results are dashed curve for displacement and un-dashed curve for velocity of the particle

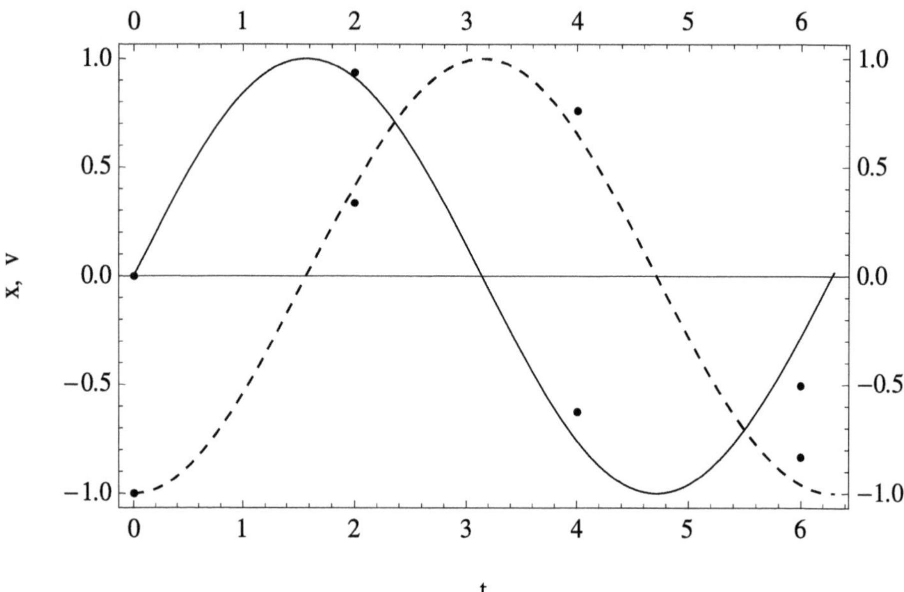

Fig. 1.9 Numerical solution of differential equation obeyed by simple harmonic motion obtained using Taylor series method using increment $h = 2.00$. Known exact results are dashed curve for displacement and un-dashed curve for velocity of the particle

```
V[i]=v=v+h*vd1+(1/2)*(h^2)*vd2+(1/6)*(h^3)*vd3+
(1/24)*(h^4)*vd4+(1/120)*(h^5)*vd5,


X[i]=x=x+h*xd1+(1/2)*(h^2)*xd2+(1/6)*(h^3)*xd3+
(1/24)*(h^4)*xd4+(1/120)*(h^5)*xd5},{t,0,6.3-h,h}];


TableForm[%,TableSpacing->{2,2},
TableHeadings->{None,{"i","t","v","x"}}]


i=-1;
p1=ListPlot[Table[{i=i+1;t=t+h,X[i]},{t,0-h,6.3-h,h}],
Frame->True,FrameLabel->{"t","x,    v"},
FrameTicks->All,PlotStyle->{Black}];


i=-1;
p2=ListPlot[Table[{i=i+1;t=t+h,V[i]},{t,0-h,6.3-h,h}],
Frame->True,FrameLabel->{"t","x,    v"},
FrameTicks->All,PlotStyle->{Black}];


p3=Plot[Sin[t],{t,0,6.3},PlotStyle->{Black}];


p4=Plot[-Cos[t],{t,0,6.3},PlotStyle->{Dashed,Black}];


Show[p1,p2,p3,p4]
```

Figures 1.2, 1.3, 1.4, 1.5, 1.6, 1.7, 1.8, and 1.9 show numerical solution of differential equation obeyed by simple harmonic motion obtained using Taylor series method using various values of increment h in the interval 0.25 to 2.00 using program number 1.2. The plot of Fig. 1.2 shows data of Table 1.2. We have obtained both velocity and displacement of the oscillating particle as functions of time. Known exact results are dashed curve for displacement and un-dashed curve for velocity of the particle executing simple harmonic motion. Agreement between numerical and analytical solutions is seen to remain perfect for increment h up to 1.00 and deteriorates for larger values of h.

1.4 Numerical Solution of Differential Equation for Damped Harmonic Motion Using Taylor Series Method

While executing damped harmonic motion, a particle is acted on by Hooke's law force $-k$ x as well as a velocity dependent damping force $-B\dfrac{dx}{dt}$, total force being

$$F = -k\,x - B\frac{dx}{dt} \tag{1.12}$$

where x is displacement of the particle from equilibrium. According to Newtonian mechanics, differential equation obeyed is

$$m\frac{d^2x}{dt^2} = -kx - B\frac{dx}{dt} \tag{1.13}$$

where m is mass of the particle, k is spring constant, B is damping constant and t is time. For simplicity, let m and k be equal in numerical value. As such differential equation (1.13) becomes

$$\frac{d^2x}{dt^2} + b\frac{dx}{dt} + x = 0 \tag{1.14}$$

Here $b = B/m$. Here we have numerically solved differential equation (1.14) as an initial value problem by Taylor series method and compared the results with known exact analytical solution given by

$$x = Ae^{-(b/2)t}\cos\left(w_d\,t + \alpha\right) \tag{1.15}$$

where

$$w_d = \sqrt{\frac{k}{m} - \left(\frac{b}{2}\right)^2} = \sqrt{1 - \left(\frac{b}{2}\right)^2} \tag{1.16}$$

Equation (1.14) is a second order differential equation whereas we need first order differential equation to deal with the Taylor series. Therefore we rewrite Eq. (1.14) as two first order differential equations as

$$\frac{dv}{dt} = -x - bv \tag{1.17}$$

and

$$\frac{dx}{dt} = v \tag{1.18}$$

To solve Eq. (1.17) using Taylor series method, iteration for numerical solution is given by

$$v_{new} = v_{old} + h\,vd1 + (1/2)h^2\,vd2 + (1/6)h^3\,vd3 + (1/24)h^4\,vd4 + (1/120)h^5\,vd5 \tag{1.19}$$

where we have denoted $\dfrac{dv}{dt}$ by $vd1$ given by $-x-bv$, $\dfrac{d^2v}{dt^2}$ by $vd2$ given by $-v-b\,vd1$, $\dfrac{d^3v}{dt^3}$ by $vd3$ given by $-vd1-b\,vd2$, $\dfrac{d^4v}{dt^4}$ by $vd4$ given by $-vd2-b\,vd3$ and $\dfrac{d^5v}{dt^5}$ by $vd5$ given by $-vd3-b\,vd4$.

To solve Eq. (1.18) using Taylor series method, iteration for numerical solution is given by

$$x_{\text{new}} = x_{\text{old}} + h\,xd1 + (1/2)h^2 xd2 + (1/6)h^3 xd3 + (1/24)h^4 xd4 + (1/120)h^5 xd5 \qquad (1.20)$$

where we have denoted $\dfrac{dx}{dt}$ by $xd1$ given by v, $\dfrac{d^2x}{dt^2}$ by $xd2$ given by $vd1$, $\dfrac{d^3x}{dt^3}$ by $xd3$ given by $vd2$, $\dfrac{d^4x}{dt^4}$ by $xd4$ given by $vd3$ and $\dfrac{d^5x}{dt^5}$ by $xd5$ given by $vd4$.

We shall explore the problem in the time interval $0 \le t \le 10$. Initial values for the problem are taken as $(t, v) = (0, 0)$ and $(t, x) = (0, \cos(\alpha))$ where $\alpha = \tan^{-1}(-b/(2\,w_d))$ obtained as follows. We have started stop watch when $v = 0$; as such we have the initial condition $(t, v) = (0, 0)$. Taking $A = 1$, Eq. (1.15) gives

$$x = e^{-(b/2)t} \cos\left(w_d\,t + \alpha\right) \qquad (1.21)$$

which gives

$$v = -(b/2)e^{-(b/2)t} \cos\left(w_d\,t + \alpha\right) - w_d\,e^{-(b/2)t} \sin\left(w_d\,t + \alpha\right)$$

At $t = 0$, we have $v = 0$ for which

$$(b/2)e^{-(b/2)t} \cos\left(w_d\,t + \alpha\right) + w_d\,e^{-(b/2)t} \sin\left(w_d\,t + \alpha\right) = 0$$

which gives

$$\alpha = \tan^{-1}\left(-\frac{b}{2w_d}\right) \qquad (1.22)$$

At $t = 0$, Eq. (1.21) gives $x = \cos(\alpha)$ where α is given by Eq. (1.22). Use of both the initial conditions is consistent with the analytic solution. The program becomes as in program number 1.3. This is symbolic computation and is evident. Resulting data are in Table 1.3 and Figs. 1.10, 1.11, 1.12, 1.13, 1.14, 1.15, 1.16, and 1.17.

Program Number 1.3 (damped harmonic motion)

```
h=0.25;
i=0;
```

Table 1.3 Numerical solution of differential equation obeyed by damped harmonic motion obtained using Taylor series method using increment $h = 0.25$ using program number 1.3. For initial values $(t, v) = (0, 0)$ and $(t, x) = (0, \cos(\alpha))$ where $\alpha = \tan^{-1}\left(-\dfrac{b}{2w_d}\right)$ where damping constant $b = 0.5$, $w_d = \sqrt{\dfrac{k}{m} - \left(\dfrac{b}{2}\right)^2} = \sqrt{1 - \left(\dfrac{b}{2}\right)^2}$

i	t	v	x
1	0.25	−0.225	0.939
2	0.50	−0.411	0.859
3	0.75	−0.550	0.738
4	1.00	−0.642	0.588
5	1.25	−0.685	0.421
6	1.50	−0.682	0.249
7	1.75	−0.641	0.083
8	2.00	−0.566	−0.068
9	2.25	−0.468	−0.198
10	2.50	−0.353	−0.301
11	2.75	−0.232	−0.374
12	3.00	−0.111	−0.417
13	3.25	0.002	−0.430
14	3.50	0.102	−0.417
15	3.75	0.184	−0.381
16	4.00	0.246	−0.327
17	4.25	0.286	−0.260
18	4.50	0.304	−0.185
19	4.75	0.303	−0.109
20	5.00	0.284	−0.035
21	5.25	0.251	0.032
22	5.50	0.207	0.089
23	5.75	0.156	0.135
24	6.00	0.102	0.167
25	6.25	0.048	0.186
26	6.50	−0.002	0.191
27	6.75	−0.046	0.185
28	7.00	−0.082	0.169
29	7.25	−0.110	0.145
30	7.50	−0.127	0.115
31	7.75	−0.135	0.082
32	8.00	−0.135	0.048
33	8.25	−0.126	0.015
34	8.50	−0.111	−0.015
35	8.75	−0.091	−0.040
36	9.00	−0.069	−0.060

(continued)

Table 1.3 (continued)

i	t	v	x
37	9.25	−0.045	−0.074
38	9.50	−0.021	−0.083
39	9.75	0.001	−0.085
40	10.00	0.021	−0.082

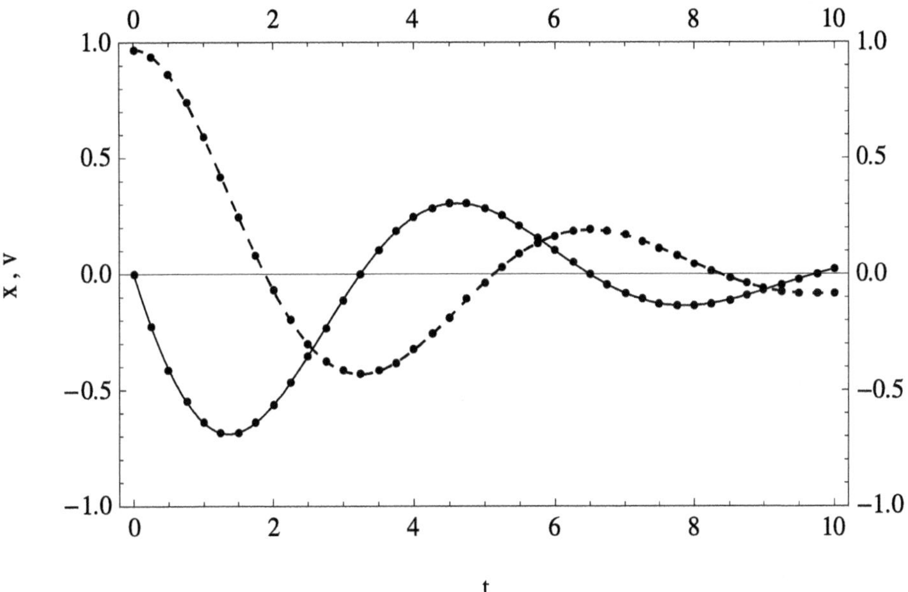

Fig. 1.10 Numerical solution of differential equation obeyed by damped harmonic motion obtained using Taylor series method using increment $h = 0.25$. Known exact results are dashed curve for displacement and un-dashed curve for velocity of the particle

```
b=0.5;
V[0]=v=0;
wd=N[Sqrt[1-(b/2)^2]];
alpha1=ArcTan[-b/(2*wd)];
X[0]=x=Cos[alpha1]

Table[{
i=i+1,
t=t+h,

vd1=-x-b*v;
vd2=-v-b*vd1;
vd3=-vd1-b*vd2;
vd4=-vd2-b*vd3;
vd5=-vd3-b*vd4;
```

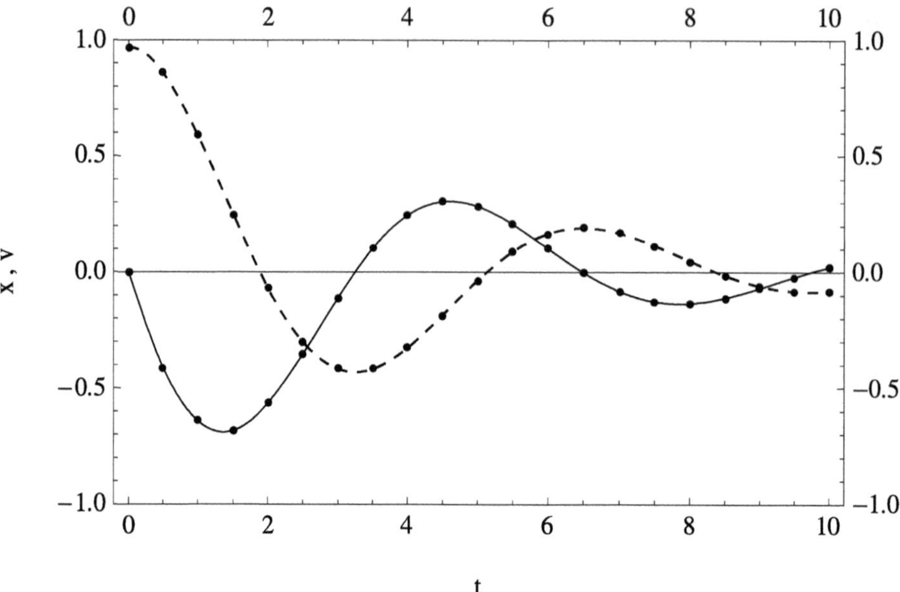

Fig. 1.11 Numerical solution of differential equation obeyed by damped harmonic motion obtained using Taylor series method using increment **h = 0.50**. Known exact results are dashed curve for displacement and un-dashed curve for velocity of the particle

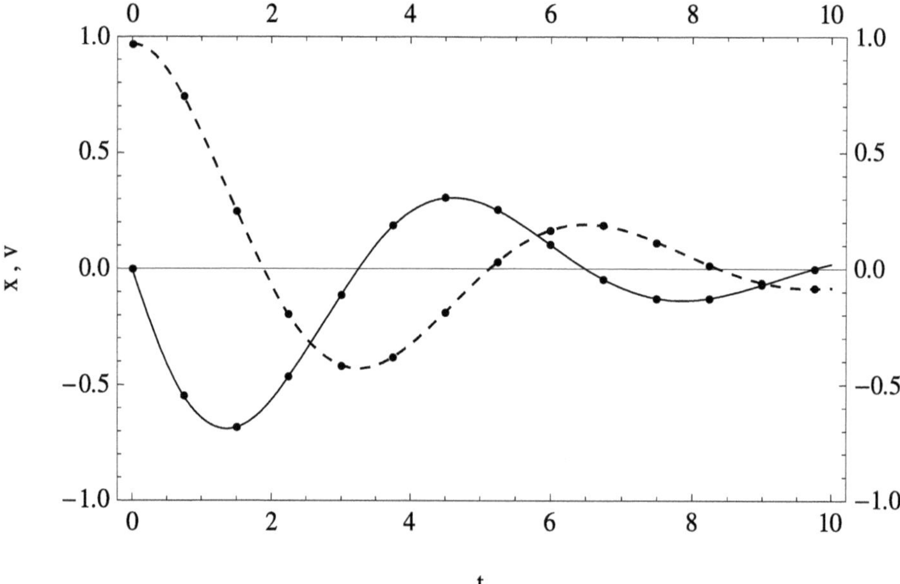

Fig. 1.12 Numerical solution of differential equation obeyed by damped harmonic motion obtained using Taylor series method using increment **h = 0.75**. Known exact results are dashed curve for displacement and un-dashed curve for velocity of the particle

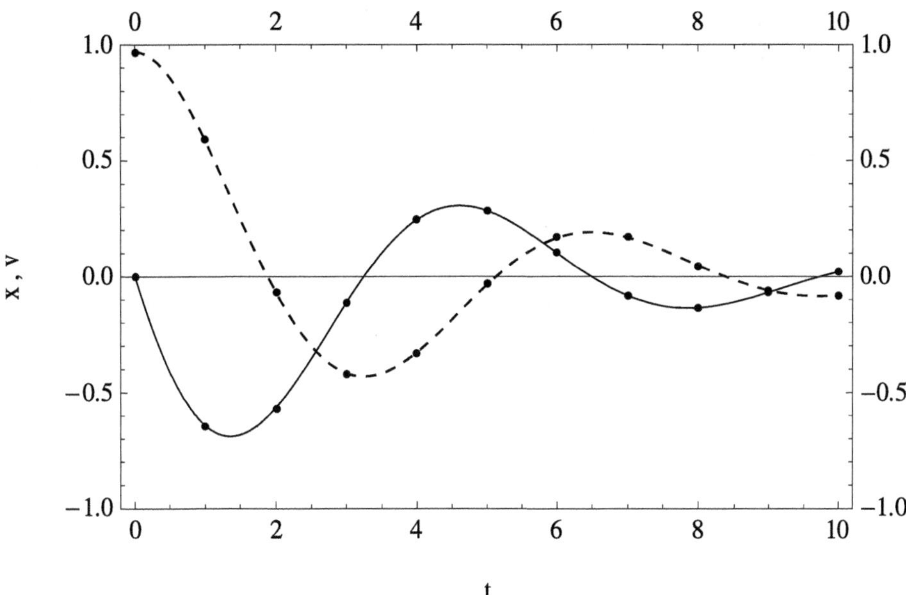

Fig. 1.13 Numerical solution of differential equation obeyed by damped harmonic motion obtained using Taylor series method using increment $h = 1.00$. Known exact results are dashed curve for displacement and un-dashed curve for velocity of the particle

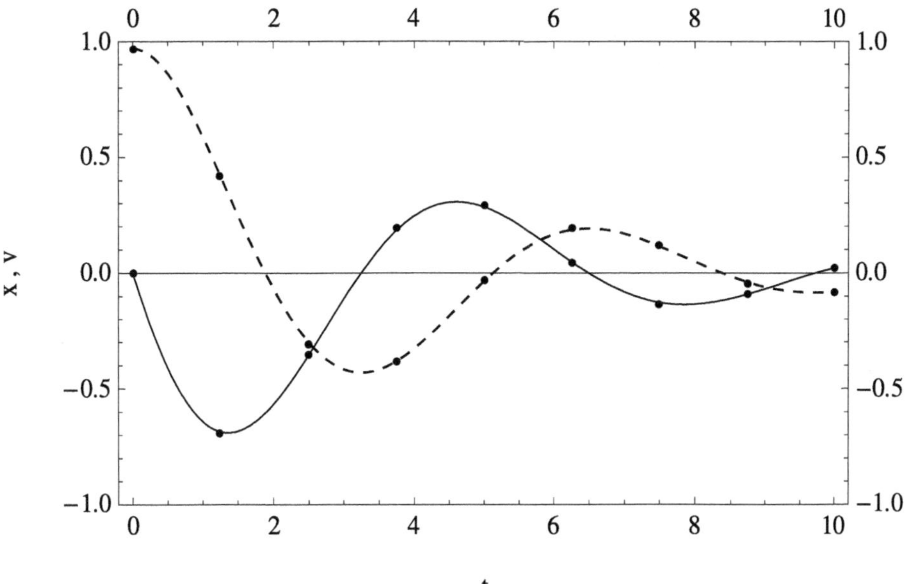

Fig. 1.14 Numerical solution of differential equation obeyed by damped harmonic motion obtained using Taylor series method using increment $h = 1.25$. Known exact results are dashed curve for displacement and un-dashed curve for velocity of the particle

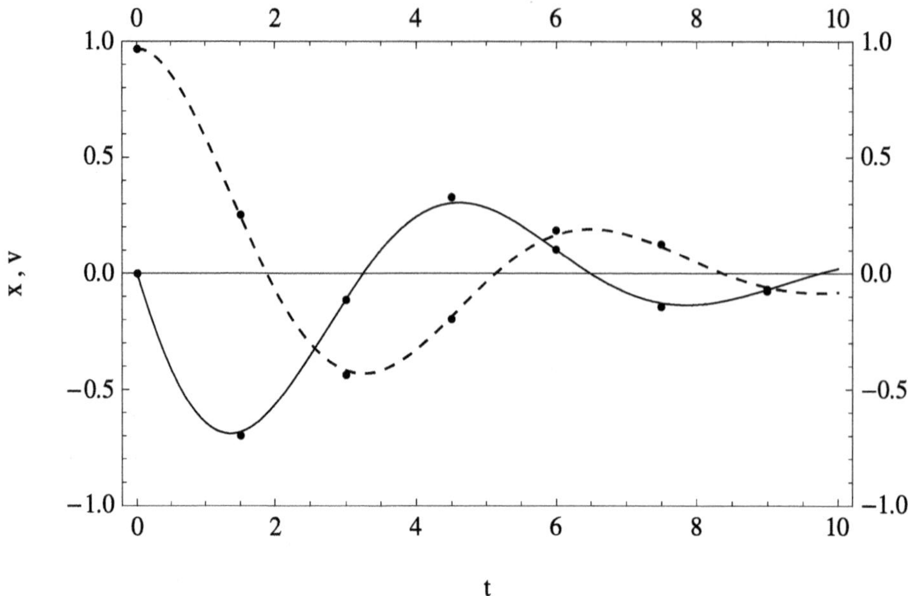

Fig. 1.15 Numerical solution of differential equation obeyed by damped harmonic motion obtained using Taylor series method using increment **h = 1.5**. Known exact results are dashed curve for displacement and un-dashed curve for velocity of the particle

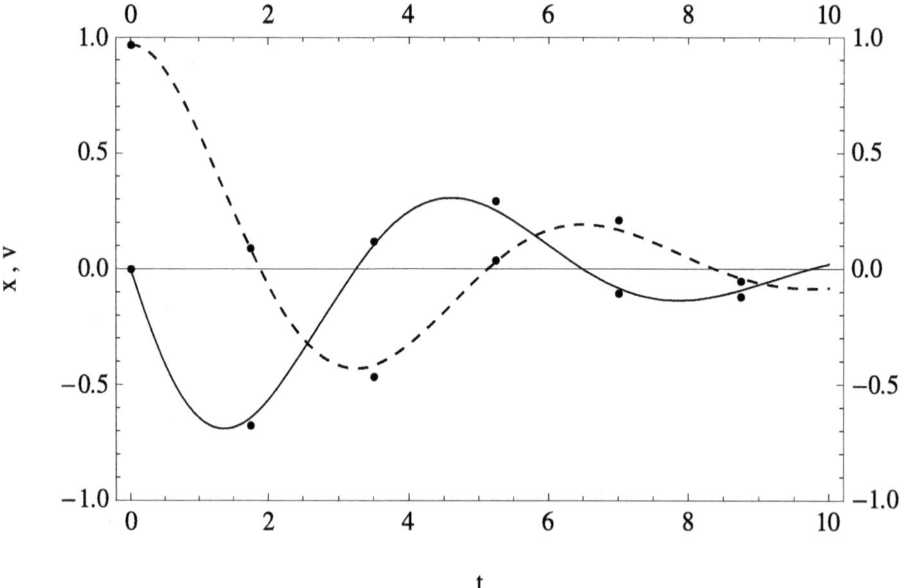

Fig. 1.16 Numerical solution of differential equation obeyed by damped harmonic motion obtained using Taylor series method using increment **h = 1.75**. Known exact results are dashed curve for displacement and un-dashed curve for velocity of the particle

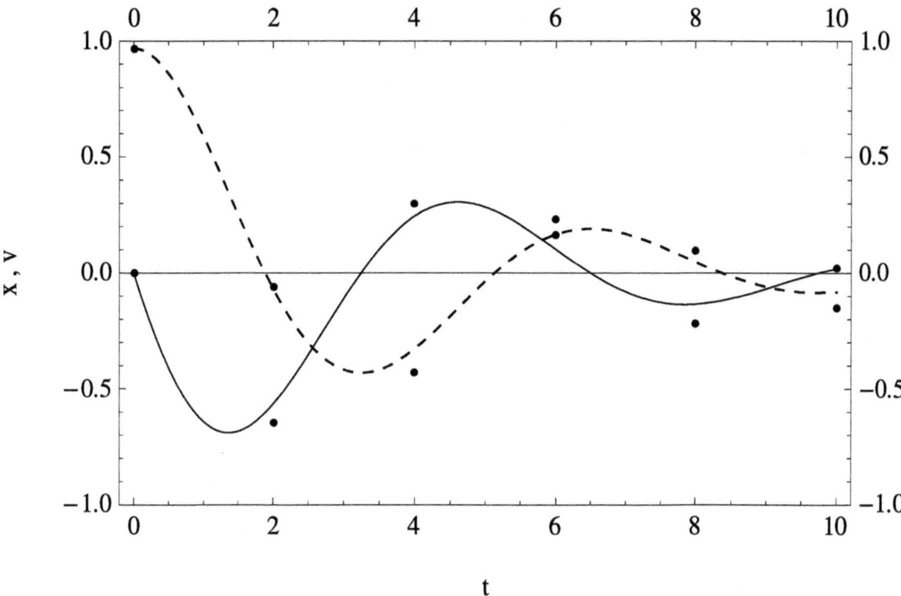

Fig. 1.17 Numerical solution of differential equation obeyed by damped harmonic motion obtained using Taylor series method using increment *h* = **2.00**. Known exact results are dashed curve for displacement and un-dashed curve for velocity of the particle

```
xd1=v;
xd2=vd1;
xd3=vd2;
xd4=vd3;
xd5=vd4;

V[i]=v=v+h*vd1+(1/2)*(h^2)*vd2+(1/6)*(h^3)*vd3+
(1/24)*(h^4)*vd4+(1/120)*(h^5)*vd5,

X[i]=x=x+h*xd1+(1/2)*(h^2)*xd2+(1/6)*(h^3)*xd3+
(1/24)*(h^4)*xd4+(1/120)*(h^5)*xd5},{t,0,10-h,h}];

TableForm[%,TableSpacing->{2,2},
TableHeadings->{None,{"i","t","v","x"}}]

i=-1;
p1=ListPlot[Table[{i=i+1;t=t+h,V[i]},{t,0-h,10-h,h}],
Frame->True,FrameLabel->{"t","x , v"},
FrameTicks->All,PlotStyle->{Black},PlotRange->{-1,1}];

i=-1;
p2=ListPlot[Table[{i=i+1;t=t+h,X[i]},{t,0-h,10-h,h}],
```

```
Frame->True,FrameLabel->{"t","x , v"},
FrameTicks->All,PlotStyle->{Black},PlotRange->{-1,1}];


p3=Plot[Exp[-(b/2)*t]*Cos[wd*t+alpha1],{t,0,10},
PlotStyle->{{Dashed,Black}},PlotRange->{-1,1}];


p4=Plot[(-b/2)*Exp[-(b/2)*t]*Cos[wd*t+alpha1]-
wd*Exp[-(b/2)*t]*Sin[wd*t+alpha1],{t,0,10},
PlotStyle->{Black},PlotRange->{-1,1}];


Show[p1,p2,p3,p4]
```

Figures 1.10, 1.11, 1.12, 1.13, 1.14, 1.15, 1.16, and 1.17 show numerical solution of differential equation obeyed by damped harmonic motion obtained using Taylor series method using various values of increment h in the interval 0.25 to 2.00, using program number 1.3. Figure 1.10 shows data of Table 1.3. We have obtained both velocity and displacement of the particle as functions of time. Known exact results are dashed curve for displacement and un-dashed curve for velocity of the particle executing damped harmonic motion. We have used initial values $(t, v) = (0, 0)$ and $(t, x) = (0, \cos(\alpha))$ where

$$\alpha = \tan^{-1}\left(-\frac{b}{2w_d}\right), \ w_d = \sqrt{\frac{k}{m}-\left(\frac{b}{2}\right)^2} = \sqrt{1-\left(\frac{b}{2}\right)^2} \ , \text{This means the particle starts with}$$

zero velocity from $x = \cos(\alpha)$. This is to make the analytic solution consistent with the initial values. Damping constant used is $b = 0.5$. Agreement between numerical and analytical solutions is seen to remain perfect for increment h up to 1.00 and deteriorates for larger values of h.

1.5   Numerical Solution of Third Order Differential Equation for Damped Harmonic Motion Using Taylor Series Method

To explore how well Taylor series method works with still higher order differential equation, we now devise a third order differential equation obeyed by damped harmonic motion. In this regard, we just take time derivative of every term of Eq. (1.14):

$$\frac{d^2 x}{dt^2}+b\frac{dx}{dt}+x = 0$$

and get

$$\frac{d^3 x}{dt^3}+b\frac{d^2 x}{dt^2}+\frac{dx}{dt} = 0 \qquad (1.23)$$

Analytical solution of differential equation (1.23) is same as that for the differential equation (1.14). In the following, we numerically solve differential equation (1.23) using Taylor series method and compare the results with the known exact analytic solution given by Eq. (1.15):

$$x = Ae^{-(b/2)t}\cos\left(w_d\, t+\alpha\right) \tag{1.24}$$

and with its time derivative which gives velocity v of the particle executing damped harmonic motion.

Equation (1.23) is a third order differential equation whereas we need first order differential equation to deal with the Taylor series. Therefore we rewrite Eq. (1.23) as three first order differential equations as

$$\frac{dw}{dt} = -v - bw \tag{1.25}$$

where

$$\frac{dv}{dt} = w \tag{1.26}$$

and

$$\frac{dx}{dt} = v \tag{1.27}$$

To solve Eq. (1.25) using Taylor series method, iteration for numerical solution is given by

$$w_{new} = w_{old} + h\,wd1 + (1/2)h^2 wd2 + (1/6)h^3 wd3 + (1/24)h^4 wd4 + (1/120)h^5 wd5 \tag{1.28}$$

where we have denoted $\dfrac{dw}{dt}$ by $wd1$ given by $-v-b\,w$, $\dfrac{d^2 w}{dt^2}$ by $wd2$ given by $-w-b\,wd1$, $\dfrac{d^3 w}{dt^3}$ by $wd3$ given by $-wd1-b\,wd2$, $\dfrac{d^4 w}{dt^4}$ by $wd4$ given by $-wd2-b\,wd3$ and $\dfrac{d^5 w}{dt^5}$ by $wd5$ given by $-wd3-b\,wd4$.

To solve Eq. (1.26) using Taylor series method, iteration for numerical solution is given by

$$v_{new} = v_{old} + h\,vd1 + (1/2)h^2 vd2 + (1/6)h^3 vd3 + (1/24)h^4 vd4 + (1/120)h^5 vd5 \tag{1.29}$$

where we have denoted $\dfrac{dv}{dt}$ by $vd1$ given by w, $\dfrac{d^2 v}{dt^2}$ by $vd2$ given by $wd1$, $\dfrac{d^3 v}{dt^3}$ by $vd3$ given by $wd2$, $\dfrac{d^4 v}{dt^4}$ by $vd4$ given by $wd3$ and $\dfrac{d^5 v}{dt^5}$ by $vd5$ given by $wd4$.

To solve Eq. (1.27) using Taylor series method, iteration for numerical solution is given by

$$x_{new} = x_{old} + h\,xd1 + (1/2)h^2 xd2 + (1/6)h^3 xd3 + (1/24)h^4 xd4 + (1/120)h^5 xd5 \qquad (1.30)$$

where we have denoted $\dfrac{dx}{dt}$ by $xd1$ given by v, $\dfrac{d^2 x}{dt^2}$ by $xd2$ given by $vd1$, $\dfrac{d^3 x}{dt^3}$ by $xd3$ given by $vd2$, $\dfrac{d^4 x}{dt^4}$ by $xd4$ given by $vd3$ and $\dfrac{d^5 x}{dt^5}$ by $xd5$ given by $vd4$.

We shall explore the problem in the time interval $0 \le t \le 10$. Initial values for the problem are taken as $(t,\ v) = (0,\ 0)$, $(t,\ x) = (0,\ \cos(\alpha))$. Initial value of w is $w = ((b/2)^2 - w_d{}^2)\cos(\alpha) + w_d\, b \sin(\alpha)$ where $\alpha = \tan^{-1}(-b/(2\,w_d))$ obtained as follows. We have started stop watch when $v = 0$; as such we have the initial condition $(t,\ v) = (0,\ 0)$. Taking $A = 1$, Eq. (1.24) gives

$$x = e^{-(b/2)t}\cos\left(w_d\,t + \alpha\right) \qquad (1.31)$$

which gives

$$v = -(b/2)e^{-(b/2)t}\cos\left(w_d t + \alpha\right) - w_d e^{-(b/2)t}\sin\left(w_d t + \alpha\right) \qquad (1.32)$$

At $t = 0$, we have $v = 0$ for which

$$(b/2)e^{-(b/2)t}\cos\left(w_d t + \alpha\right) + w_d\, e^{-(b/2)t}\sin\left(w_d t + \alpha\right) = 0$$

which gives

$$\alpha = \tan^{-1}\left(-\frac{b}{2w_d}\right) \qquad (1.33)$$

At $t = 0$, Eq. (1.31) gives $x = \cos(\alpha)$ where α is given by Eq. (1.33). Time derivative of v obtained from Eq. (1.32) is

$$w = (b/2)^2\, e^{-(b/2)t}\cos\left(w_d t + \alpha\right) + w_d\,(b/2)e^{-(b/2)t}\sin\left(w_d t + \alpha\right)$$
$$+ w_d\,(b/2)e^{-(b/2)t}\sin\left(w_d t + \alpha\right) - w_d^2\, e^{-(b/2)t}\cos\left(w_d t + \alpha\right)$$

At $t = 0$, we have

$$w = (b/2)^2\cos(\alpha) + w_d\,(b/2)\sin(\alpha) + w_d\,(b/2)\sin(\alpha) - w_d^2\cos(\alpha) \qquad (1.34)$$

This is initial value of w. Use of all the three initial conditions is consistent with the analytic solution. The program becomes as in program number 1.4. This is symbolic computation and is evident. Resulting data are in Table 1.4 and in Figs. 1.18, 1.19, 1.20, 1.21, 1.22, 1.23, and 1.24.

Program Number 1.4 (third order differential equation for damped harmonic motion)

Table 1.4 Numerical solution of third order differential equation obeyed by damped harmonic motion obtained using Taylor series method using increment $h = 0.25$ using program number 1.4. For initial values $(t, w) = (0, (b/2)^2 \cos(\alpha) + w_d (b/2)\sin(\alpha) + w_d (b/2)\sin(\alpha) - w_d^2 \cos(\alpha))$, $(t, v) = (0, 0)$ and $(t, x) = (0, \cos(\alpha))$ where $\alpha = \tan^{-1}\left(-\dfrac{b}{2w_d}\right)$ where damping constant $b = 0.5$, $w_d = \sqrt{\dfrac{k}{m} - \left(\dfrac{b}{2}\right)^2} = \sqrt{1 - \left(\dfrac{b}{2}\right)^2}$

i	t	w	v	x
1	0.25	−0.827	−0.225	0.939
2	0.50	−0.654	−0.411	0.859
3	0.75	−0.463	−0.550	0.738
4	1.00	−0.267	−0.642	0.588
5	1.25	−0.079	−0.685	0.421
6	1.50	0.092	−0.682	0.249
7	1.75	0.237	−0.641	0.083
8	2.00	0.352	−0.566	−0.068
9	2.25	0.432	−0.468	−0.198
10	2.50	0.478	−0.353	−0.301
11	2.75	0.490	−0.232	−0.374
12	3.00	0.472	−0.111	−0.417
13	3.25	0.429	0.002	−0.430
14	3.50	0.366	0.102	−0.417
15	3.75	0.289	0.184	−0.381
16	4.00	0.204	0.246	−0.327
17	4.25	0.117	0.286	−0.260
18	4.50	0.033	0.304	−0.185
19	4.75	−0.042	0.303	−0.109
20	5.00	−0.107	0.284	−0.035
21	5.25	−0.157	0.251	0.032
22	5.50	−0.193	0.207	0.089
23	5.75	−0.213	0.156	0.135
24	6.00	−0.218	0.102	0.167
25	6.25	−0.210	0.048	0.186
26	6.50	−0.190	−0.002	0.191
27	6.75	−0.162	−0.046	0.185
28	7.00	−0.127	−0.082	0.169
29	7.25	−0.090	−0.110	0.145
30	7.50	−0.051	−0.127	0.115
31	7.75	−0.014	−0.135	0.082
32	8.00	0.020	−0.135	0.048
33	8.25	0.048	−0.126	0.015
34	8.50	0.070	−0.111	−0.015
35	8.75	0.086	−0.091	−0.040

(continued)

Table 1.4 (continued)

i	t	w	v	x
36	9.00	0.095	−0.069	−0.060
37	9.25	0.097	−0.045	−0.074
38	9.50	0.093	−0.021	−0.083
39	9.75	0.084	0.001	−0.085
40	10.00	0.072	0.021	−0.082

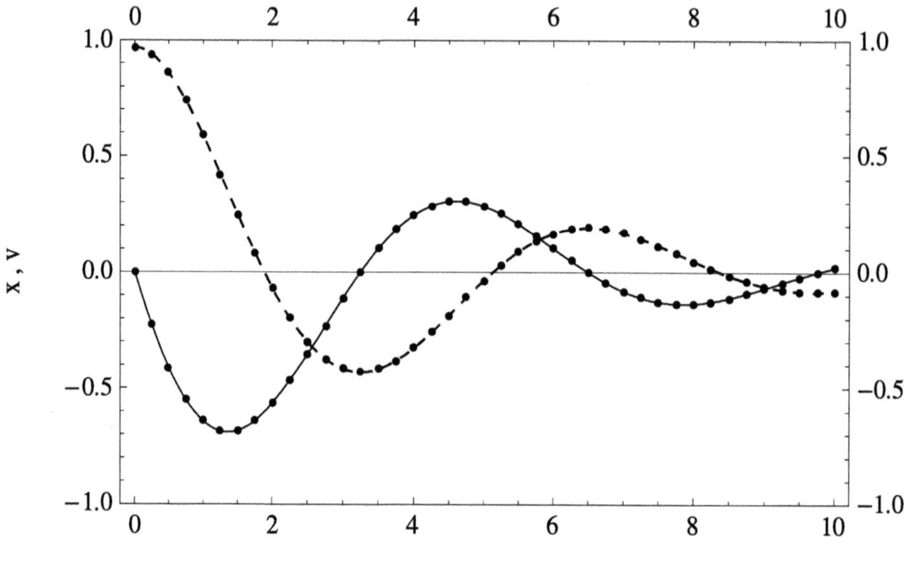

Fig. 1.18 Numerical solution of third order differential equation obeyed by damped harmonic motion obtained using Taylor series method using increment **$h = 0.25$**. Known exact results are dashed curve for displacement and un-dashed curve for velocity of the particle

```
h=0.25;
i=0;
b=0.5;
V[0]=v=0;
wd=N[Sqrt[1-(b/2)^2]];
alpha1=ArcTan[-b/(2*wd)];
X[0]=x=Cos[alpha1]
W[0]=w=((b/2)^2-wd^2)*Cos[alpha1]+wd*b*Sin[alpha1];

Table[{
i=i+1,
t=t+h,
```

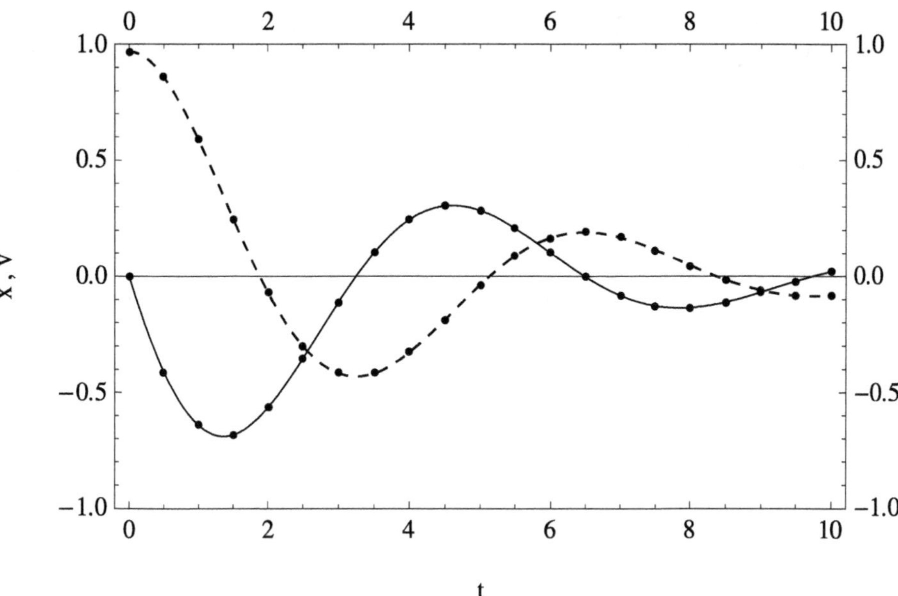

Fig. 1.19 Numerical solution of third order differential equation obeyed by damped harmonic motion obtained using Taylor series method using increment $h = 0.5$. Known exact results are dashed curve for displacement and un-dashed curve for velocity of the particle

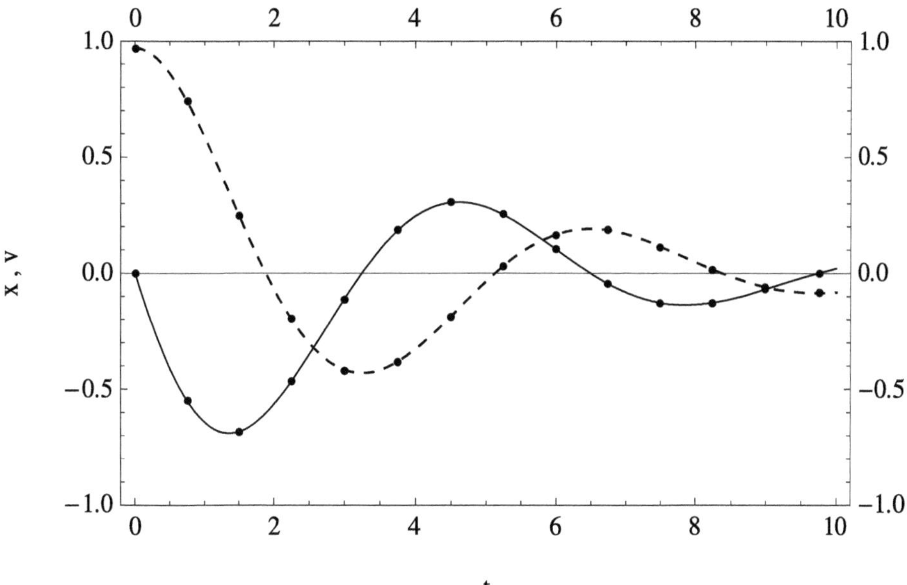

Fig. 1.20 Numerical solution of third order differential equation obeyed by damped harmonic motion obtained using Taylor series method using increment $h = 0.75$. Known exact results are dashed curve for displacement and un-dashed curve for velocity of the particle

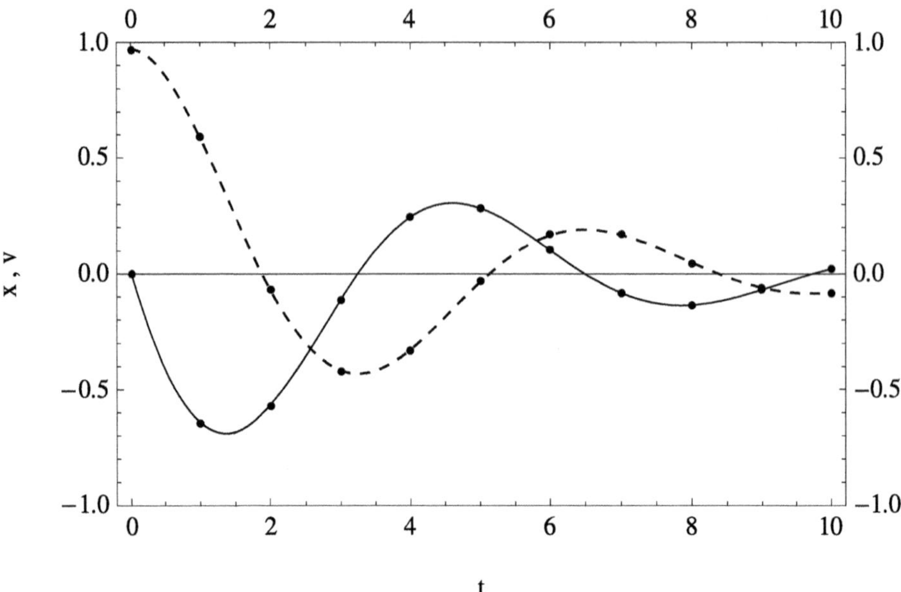

Fig. 1.21 Numerical solution of third order differential equation obeyed by damped harmonic motion obtained using Taylor series method using increment $h = 1.0$. Known exact results are dashed curve for displacement and un-dashed curve for velocity of the particle

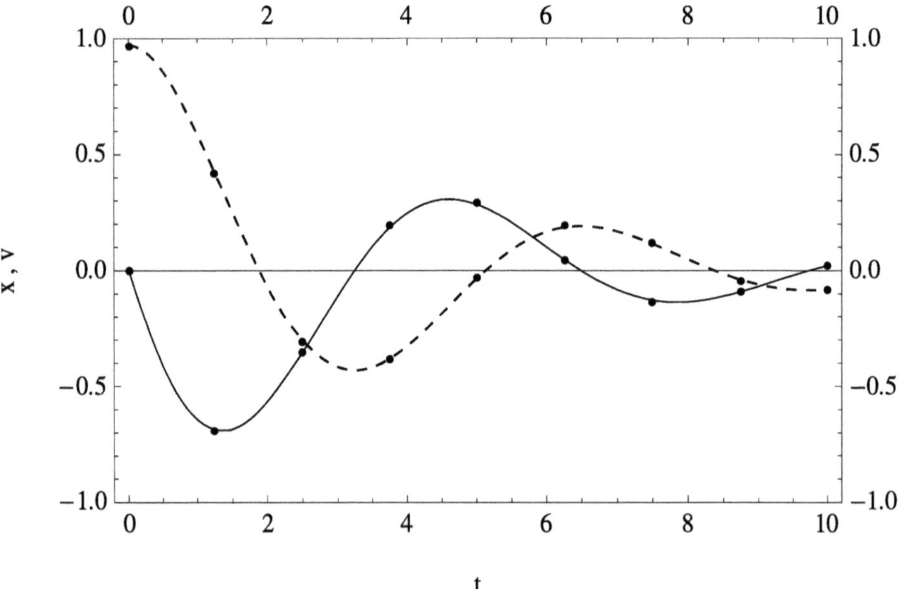

Fig. 1.22 Numerical solution of third order differential equation obeyed by damped harmonic motion obtained using Taylor series method using increment $h = 1.25$. Known exact results are dashed curve for displacement and un-dashed curve for velocity of the particle

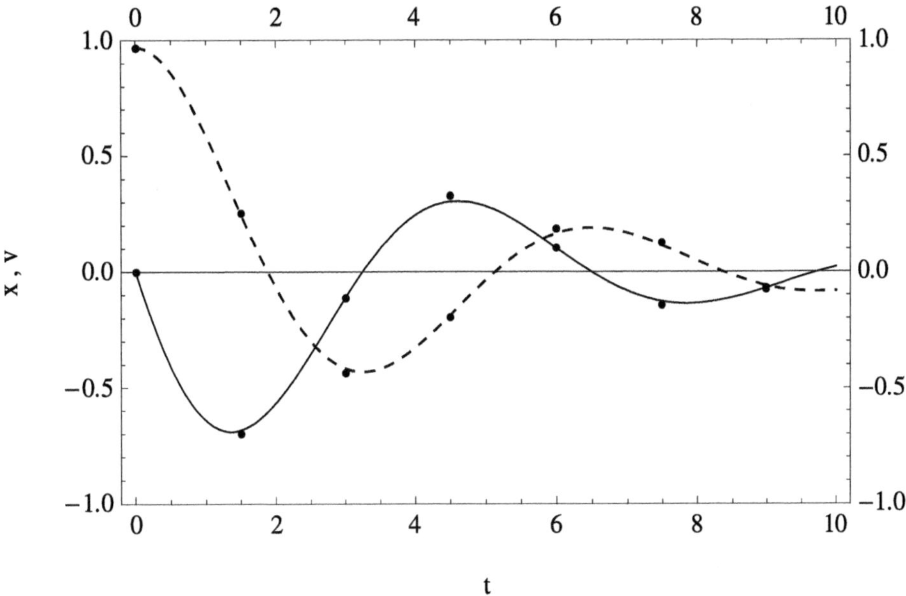

Fig. 1.23 Numerical solution of third order differential equation obeyed by damped harmonic motion obtained using Taylor series method using increment $h = 1.5$. Known exact results are dashed curve for displacement and un-dashed curve for velocity of the particle

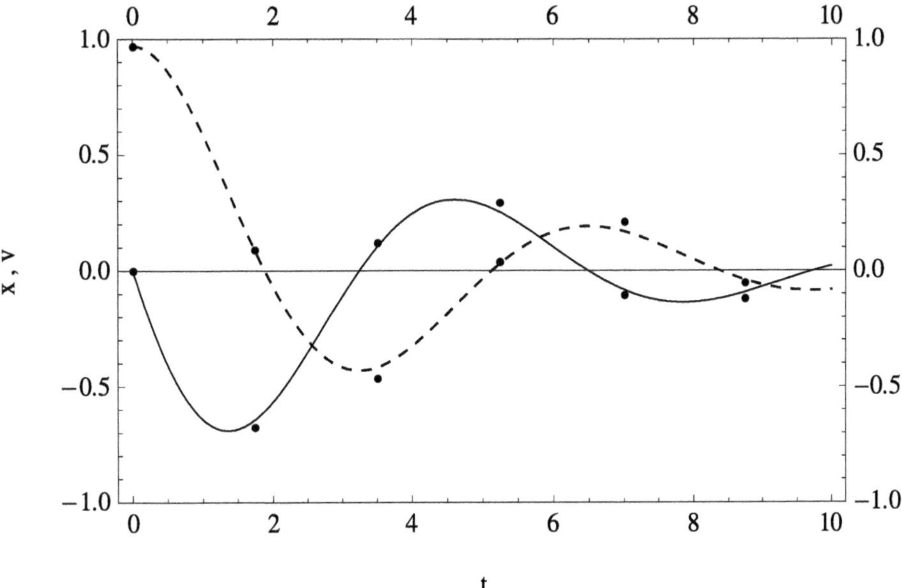

Fig. 1.24 Numerical solution of third order differential equation obeyed by damped harmonic motion obtained using Taylor series method using increment $h = 1.75$. Known exact results are dashed curve for displacement and un-dashed curve for velocity of the particle

```
wd1=-v-b*w;
wd2=-w-b*wd1;
wd3=-wd1-b*wd2;
wd4=-wd2-b*wd3;
wd5=-wd3-b*wd4;

vd1=w;
vd2=wd1;
vd3=wd2;
vd4=wd3;
vd5=wd4;

xd1=v;
xd2=vd1;
xd3=vd2;
xd4=vd3;
xd5=vd4;

W[i]=w=w+h*wd1+(1/2)*(h^2)*wd2+(1/6)*(h^3)*wd3+
(1/24)*(h^4)*wd4+(1/120)*(h^5)*wd5,

V[i]=v=v+h*vd1+(1/2)*(h^2)*vd2+(1/6)*(h^3)*vd3+
(1/24)*(h^4)*vd4+(1/120)*(h^5)*vd5,

X[i]=x=x+h*xd1+(1/2)*(h^2)*xd2+(1/6)*(h^3)*xd3+
(1/24)*(h^4)*xd4+(1/120)*(h^5)*xd5},{t,0,10-h,h}];

TableForm[%,TableSpacing->{2,2},
TableHeadings->{None,{"i","t","w","v","x"}}]

i=-1;
p1=ListPlot[Table[{i=i+1;t=t+h,V[i]},{t,0-h,10-h,h}],
Frame->True,FrameLabel->{"t","x , v"},
FrameTicks->All,PlotStyle->{Black},PlotRange->{-1,1}];

i=-1;
p2=ListPlot[Table[{i=i+1;t=t+h,X[i]},{t,0-h,10-h,h}],
Frame->True,FrameLabel->{"t","x , v"},
FrameTicks->All,PlotStyle->{Black},PlotRange->{-1,1}];

p3=Plot[Exp[-(b/2)*t]*Cos[wd*t+alpha1],{t,0,10},
PlotStyle->{{Dashed,Black}},PlotRange->{-1,1}];

p4=Plot[(-b/2)*Exp[-(b/2)*t]*Cos[wd*t+alpha1]-
wd*Exp[-(b/2)*t]*Sin[wd*t+alpha1],{t,0,10},
```

```
PlotStyle->{Black},PlotRange->{-1,1}];
```

```
Show[p1,p2,p3,p4]
```

Figures 1.18, 1.19, 1.20, 1.21, 1.22, 1.23, and 1.24 show numerical solution of third order differential equation obeyed by damped harmonic motion obtained using Taylor series method using various values of increment h in the interval 0.25 to 1.75, using program number 1.4. Figure 1.18 shows data of Table 1.4. Known exact results are dashed curve for displacement and un-dashed curve for velocity of the particle executing damped harmonic motion. For initial values $(t,\ w)\ =\ (0,\ (b/2)^2\cos(\alpha)+w_d(b/2)\sin(\alpha)+w_d(b/2)\sin(\alpha)-w_d^2\cos(\alpha))$, $(t,\ v) = (0,\ 0)$ and $(t,\ x)\ =\ (0,\ \cos(\alpha))$ which are consistent with the analytical solution used. Here

$$\alpha = \tan^{-1}\left(-\frac{b}{2w_d}\right),\ w_d = \sqrt{\frac{k}{m}-\left(\frac{b}{2}\right)^2} = \sqrt{1-\left(\frac{b}{2}\right)^2}\ ,\ \text{and damping constant } b = 0.5.$$

We have numerically solved Newton's differential equation of motion obeyed by damped harmonic motion as an initial value problem using Taylor series method. We first turned the differential equation into a third order differential equation. This is intended to explore and check if Taylor series method works well with third order differential equation. We find that we can use large values of increment of the independent variable (time) and get numerical solutions that closely match with known exact analytic solution. We have obtained both velocity and displacement of the oscillating particle as functions of time. The agreement remains perfect for increment h up to 1.25 and deteriorates for larger values of h.

1.6 Numerical Solution of Differential Equation for Driven Damped Harmonic Motion Using Taylor Series Method

While executing driven damped harmonic oscillation, a particle is acted on by Hooke's law force $-kx$ as well as a velocity dependent damping force $-B\dfrac{dx}{dt}$, besides an oscillatory driving force $F_{max}\cos(\omega\ t)$ e.g., total force being

$$F = -k\ x - B\frac{dx}{dt} + F_{max}\cos(\omega\ t) \tag{1.35}$$

where x is displacement of the particle from equilibrium. According to Newtonian mechanics, differential equation obeyed during the motion is

$$m\frac{d^2x}{dt^2} = -kx - B\frac{dx}{dt} + F_{max}\cos(\omega t) \tag{1.36}$$

where m is mass of the particle, k is spring constant, B is damping constant, F_{max} is maximum value of oscillatory driving force of angular frequency ω, and t is time. The differential equation can be written as

$$\frac{d^2 x}{dt^2} + b\frac{dx}{dt} + \omega_0^2 x = a\cos(\omega t) \tag{1.37}$$

Here $b = B/m$, $\omega_0^2 = \dfrac{k}{m}$, $a = F_{max}/m$. Here we have numerically solved differential equation (1.37) as an initial value problem by Taylor series method and compared the results with known analytic solution:

$$x = Ae^{-(b/2)t}\cos(w_d\, t + \alpha) + \frac{a}{\sqrt{\left(\omega^2 - \omega_0^2\right)^2 + b^2\omega^2}}\sin\left[\omega t - \tan^{-1}\left(\frac{\omega^2 - \omega_0^2}{b\omega}\right)\right] \tag{1.38}$$

where

$$w_d = \sqrt{\frac{k}{m} - \left(\frac{b}{2}\right)^2} \tag{1.39}$$

But Eq. (1.37) is a second order differential equation whereas we need first order differential equation to deal with the Taylor series. Therefore we rewrite Eq. (1.37) as two first order differential equations as:

$$\frac{dv}{dt} = -bv - \omega_0^2 x + a\cos(\omega t) \tag{1.40}$$

and

$$\frac{dx}{dt} = v \tag{1.41}$$

To solve Eq. (1.40) using Taylor series method, iteration for numerical solution is given by

$$v_{new} = v_{old} + h\, vd1 + (1/2)h^2 vd2 + (1/6)h^3 vd3 + (1/24)h^4 vd4 + (1/120)h^5 vd5 \tag{1.42}$$

where we have denoted $\dfrac{dv}{dt}$ by $vd1$, $\dfrac{d^2 v}{dt^2}$ by $vd2$, $\dfrac{d^3 v}{dt^3}$ by $vd3$, $\dfrac{d^4 v}{dt^4}$ by $vd4$ and $\dfrac{d^5 v}{dt^5}$ by $vd5$.

To solve Eq. (1.41) using Taylor series method, iteration for numerical solution is given by

$$x_{new} = x_{old} + h\, xd1 + (1/2)h^2 xd2 + (1/6)h^3 xd3 + (1/24)h^4 xd4 + (1/120)h^5 xd5 \tag{1.43}$$

where we have denoted $\dfrac{dx}{dt}$ by $xd1$, $\dfrac{d^2x}{dt^2}$ by $xd2$, $\dfrac{d^3x}{dt^3}$ by $xd3$, $\dfrac{d^4x}{dt^4}$ by $xd4$ and $\dfrac{d^5x}{dt^5}$ by $xd5$.

We shall explore the problem in the time interval $0 \le t \le 60$. Initial values of v and of x for the problem are taken using Eq. (1.38) as

$$(t,v) = \left(0{,}A(-b/2)\cos(\alpha) - A\omega_d \sin(\alpha) + \frac{a\omega}{\sqrt{\left(\omega^2 - \omega_0^2\right)^2 + b^2\omega^2}} \cos\left[-\tan^{-1}\left(\frac{\omega^2 - \omega_0^2}{b\omega} \right) \right] \right)$$

and

$$(t,x) = \left(0{,}A\cos(\alpha) + \frac{a}{\sqrt{\left(\omega^2 - \omega_0^2\right)^2 + b^2\omega^2}} \sin\left[-\tan^{-1}\left(\frac{\omega^2 - \omega_0^2}{b\omega} \right) \right] \right)$$

The program becomes as in program number 1.5. This is symbolic computation and is evident. Resulting data are in Table 1.5 and in Figs. 1.25, 1.26, 1.27, 1.28, 1.29, 1.30, and 1.31.

Program Number 1.5 (driven damped harmonic motion)

```
T=10;
T0=5;

w=2*Pi/T;
w0=2*Pi/T0;

A=5;
a=1;
b=0.2;
h=0.5;

i=0;
wd=Sqrt[w0^2-(b/2)^2];
Td=2*Pi/wd
alpha1=0;

X[0]=x=A*Cos[alpha1]+(a/Sqrt[(w0^2-w^2)^2+
(b*w)^2])*Sin[-ArcTan[(w^2-w0^2)/(b*w)]]

V[0]=v=A*(-b/2)*Cos[alpha1]-A*wd*Sin[alpha1]+
(a/Sqrt[(w0^2-w^2)^2+(b*w)^2])*w*
Cos[-ArcTan[(w^2-w0^2)/(b*w)]]
```

Table 1.5 Numerical solution of differential equation (1.37): $\dfrac{d^2x}{dt^2} + b\dfrac{dx}{dt} + \omega_0^2 x = a\cos(\omega t)$ obeyed by driven damped harmonic motion obtained using Taylor series method using increment $h = 0.5$ using program number 1.5. For parameters as in program 1.5

i	t	x (approx.)
1	0.5	4.675
2	1.0	2.143
3	1.5	−0.743
4	2.0	−2.951
5	2.5	−3.807
6	3.0	−3.198
7	3.5	−1.555
8	4.0	0.362
9	4.5	1.780
10	5.0	2.200
11	5.5	1.551
12	6.0	0.184
13	6.5	−1.305
14	7.0	−2.312
15	7.5	−2.453
16	8.0	−1.688
17	8.5	−0.311
18	9.0	1.183
19	9.5	2.289
20	10.0	2.676
21	10.5	2.282
22	11.0	1.312
23	11.5	0.142
24	12.0	−0.835
25	12.5	−1.345
26	13.0	−1.319
27	13.5	−0.887
28	14.0	−0.308
29	14.5	0.143
30	15.0	0.282
31	15.5	0.077
32	16.0	−0.354
33	16.5	−0.799
34	17.0	−1.045
35	17.5	−0.958
36	18.0	−0.530
37	18.5	0.121
38	19.0	0.800
39	19.5	1.310
40	20.0	1.512

(continued)

Table 1.5 (continued)

i	t	x (approx.)
41	20.5	1.373
42	21.0	0.966
43	21.5	0.430
44	22.0	−0.077
45	22.5	−0.438
46	23.0	−0.606
47	23.5	−0.609
48	24.0	−0.525
49	24.5	−0.443
50	25.0	−0.425
51	25.5	−0.482
52	26.0	−0.576
53	26.5	−0.636
54	27.0	−0.592
55	27.5	−0.408
56	28.0	−0.092
57	28.5	0.298
58	29.0	0.678
59	29.5	0.960
60	30.0	1.083
61	30.5	1.030
62	31.0	0.824
63	31.5	0.522
64	32.0	0.193
65	32.5	−0.105
66	33.0	−0.337
67	33.5	−0.496
68	34.0	−0.594
69	34.5	−0.652
70	35.0	−0.685
71	35.5	−0.694
72	36.0	−0.667
73	36.5	−0.584
74	37.0	−0.431
75	37.5	−0.206
76	38.0	0.074
77	38.5	0.371
78	39.0	0.639
79	39.5	0.835
80	40.0	0.926
81	40.5	0.900
82	41.0	0.766
83	41.5	0.551

(continued)

Table 1.5 (continued)

i	t	x (approx.)
84	42.0	0.289
85	42.5	0.018
86	43.0	−0.235
87	43.5	−0.450
88	44.0	−0.615
89	44.5	−0.726
90	45.0	−0.780
91	45.5	−0.773
92	46.0	−0.703
93	46.5	−0.569
94	47.0	−0.374
95	47.5	−0.131
96	48.0	0.136
97	48.5	0.400
98	49.0	0.628
99	49.5	0.791
100	50.0	0.868
101	50.5	0.851
102	51.0	0.743
103	51.5	0.559
104	52.0	0.324
105	52.5	0.063
106	53.0	−0.197
107	53.5	−0.431
108	54.0	−0.621
109	54.5	−0.752
110	55.0	−0.815
111	55.5	−0.803
112	56.0	−0.718
113	56.5	−0.564
114	57.0	−0.353
115	57.5	−0.104
116	58.0	0.160
117	58.5	0.412
118	59.0	0.624
119	59.5	0.775
120	60.0	0.847

```
Table[{
i=i+1,
t=t+h,
```

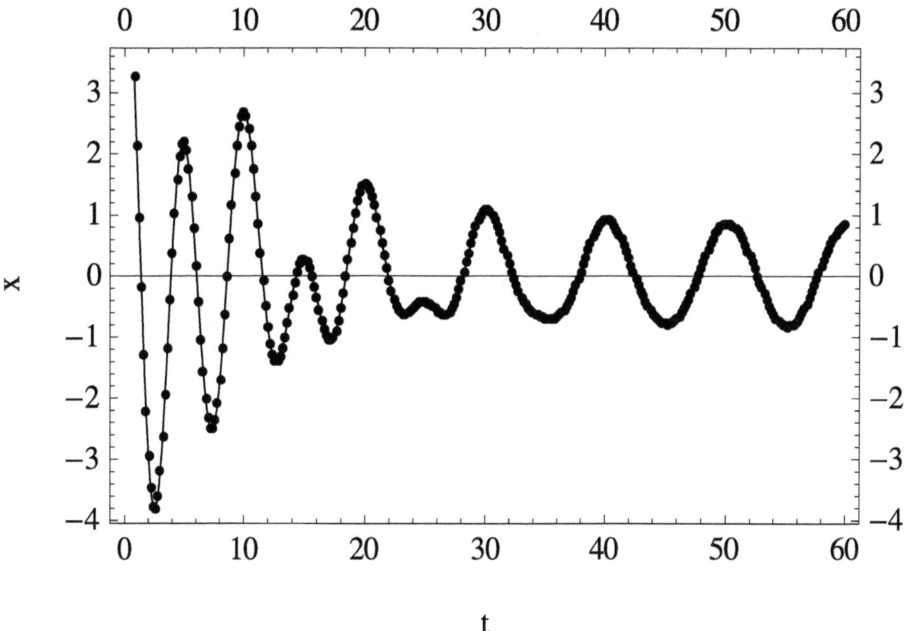

Fig. 1.25 Numerical solution of differential equation obeyed by driven damped harmonic oscillation obtained using Taylor series method using increment **h = 0.2**, showing displacement of the particle as function of time.. The curve is analytic solution

```
vd1=-b*v-(w0^2)*x+a*Cos[w*(t-h)];
vd2=-b*vd1-(w0^2)*v-w*a*Sin[w*(t-h)];
vd3=-b*vd2-(w0^2)*vd1-(w^2)*a*Cos[w*(t-h)];
vd4=-b*vd3-(w0^2)*vd2+(w^3)*a*Sin[w*(t-h)];
vd5=-b*vd4-(w0^2)*vd3+(w^4)*a*Cos[w*(t-h)];

xd1=v;
xd2=vd1;
xd3=vd2;
xd4=vd3;
xd5=vd4;

V[i]=v=v+h*vd1+(1/2)*h^2*vd2+(1/6)*h^3*vd3+
(1/24)*h^4*vd4+(1/120)*h^5*vd5;

X[i]=x=x+h*xd1+(1/2)*h^2*xd2+(1/6)*h^3*xd3+
(1/24)*h^4*xd4+(1/120)*h^5*xd5},

{t,0,60-h,h}];
TableForm[%,TableSpacing->{2,2},
TableHeadings->{None,{"i","t","x (approx.)"}}]
```

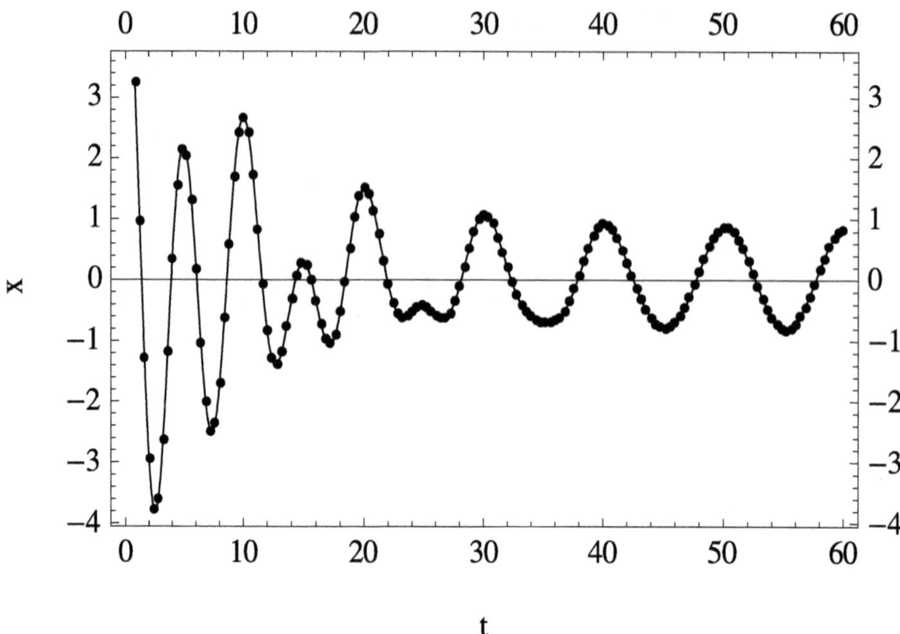

Fig. 1.26 Numerical solution of differential equation obeyed by driven damped harmonic oscillation obtained using Taylor series method using increment **h = 0.4**, showing displacement of the particle as function of time.. The curve is analytic solution

```
i=-1;
p1=ListPlot[Table[{i=i+1;t=t+h,X[i]},{t,0-h,60-h,h}],
Frame->True,
FrameLabel->{"t","x"},FrameTicks->All,
PlotStyle->{Black}];

p2=Plot[(A*Exp[-b*t/2])*Cos[wd*t]+
(a/Sqrt[(w0^2-w^2)^2+(b*w)^2])*
Sin[w*t-ArcTan[(w^2-w0^2)/(b*w)]],
{t,0,60},PlotStyle->{Black}];

Show[p1,p2]
```

Figures 1.25, 1.26, 1.27, 1.28, 1.29, 1.30, and 1.31 show numerical solution of differential equation obeyed by driven damped harmonic oscillation obtained using Taylor series method using various values of increment h in the interval 0.2 to 1.4, using program number 1.5. The curve is analytic solution and dots are numerical solution showing displacement of the particle executing driven damped harmonic oscillations as function of time t. For initial values indicated in program number 1.5. We find that agreement remains perfect for increment up to 1.00 and deteriorates for larger values of h.

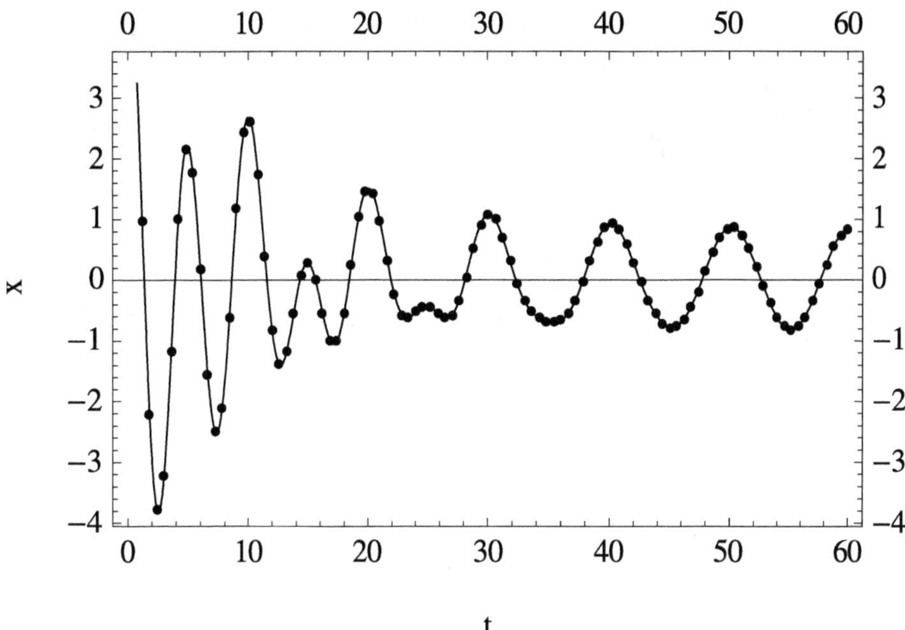

Fig. 1.27 Numerical solution of differential equation obeyed by driven damped harmonic oscillation obtained using Taylor series method using increment **h = 0.6**, showing displacement of the particle as function of time.. The curve is analytic solution

We have numerically solved Newton's differential equation of motion obeyed by driven damped harmonic oscillation as initial value problem using Taylor series method. We find that we can use large values of increments of the independent variable (time) and get numerical solutions that closely match with known exact analytic solution. We have superposition of oscillations of two different frequencies; it fades away with time and steady oscillation of frequency of driving force persists for large values of time.

1.7   Motion of Oscillators in Phase Space Obtained Using Taylor Series Method

We now modify program number 1.2 to obtain speed v versus displacement x plot of a particle executing simple harmonic motion. The modified program is program number 1.6. The motion in phase space is as in Fig. 1.32. We find that the particle regains same speed periodically.

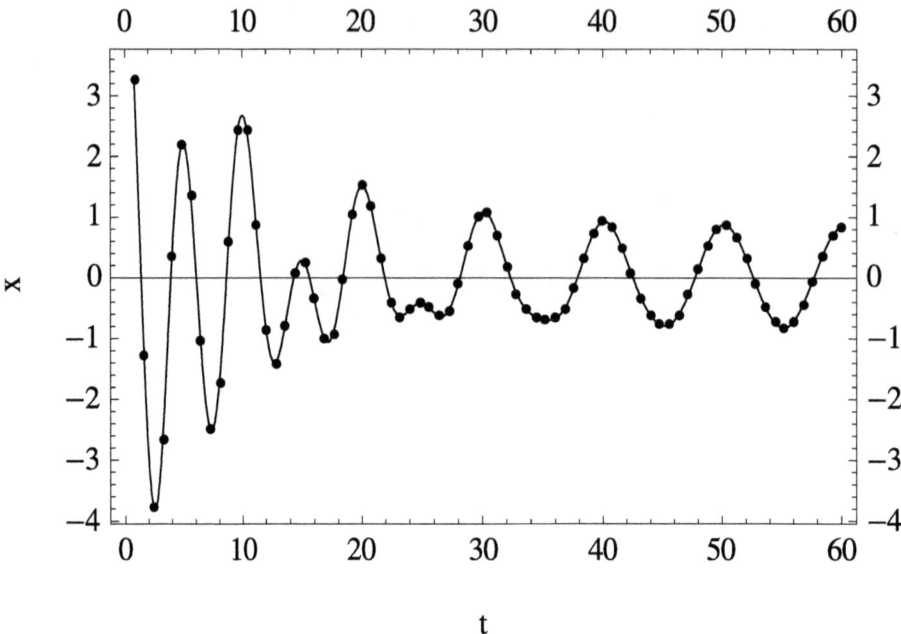

Fig. 1.28 Numerical solution of differential equation obeyed by driven damped harmonic oscilla-tion obtained using Taylor series method using increment $h = 0.8$, showing displacement of the particle as function of time.. The curve is analytic solution

Program Number 1.6 (simple harmonic motion in phase space)

```
h=0.1;
X[0]=x=-1;
V[0]=v=0;
i=0;

Table[{
i=i+1,
t=t+h,

vd1=-x;
vd2=-v;
vd3=x;
vd4=v;
vd5=-x;

xd1=v;
xd2=vd1;
xd3=vd2;
xd4=vd3;
xd5=vd4;
```

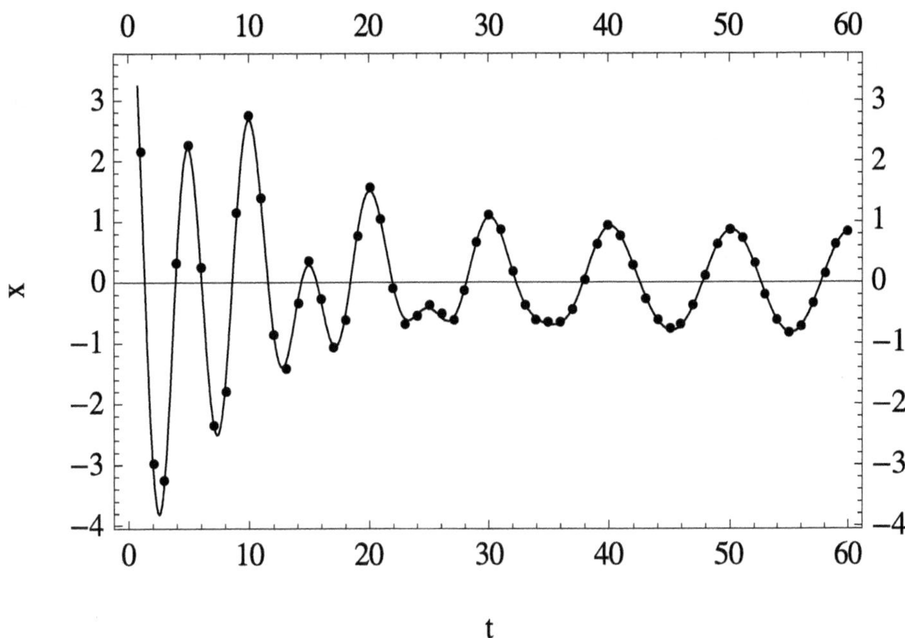

Fig. 1.29 Numerical solution of differential equation obeyed by driven damped harmonic oscillation obtained using Taylor series method using increment **h = 1.0**, showing displacement of the particle as function of time.. The curve is analytic solution

```
V[i]=v=v+h*vd1+(1/2)*(h^2)*vd2+(1/6)*(h^3)*vd3+
(1/24)*(h^4)*vd4+(1/120)*(h^5)*vd5,

X[i]=x=x+h*xd1+(1/2)*(h^2)*xd2+(1/6)*(h^3)*xd3+
(1/24)*(h^4)*xd4+(1/120)*(h^5)*xd5},

{t,0,6.3-h,h}];
TableForm[%,TableSpacing->{2,2},
TableHeadings->{None,{"i","t","v","x"}}]

i=-1;
p1=ListLinePlot[Table[{i=i+1;t=t+h;X[i],V[i]},{t,0-h,6.3-h,h}],
Frame->True,FrameLabel->{"x","v"},FrameTicks->All,
PlotStyle->{Black,Thick},AspectRatio->Automatic]
```

We now modify program number 1.3 to obtain speed v versus displacement x plot of a particle executing damped harmonic motion. The modified program is program number 1.7. The motion in phase space is as in Fig. 1.33. We find that speed of the particle continues to decrease as oscillation dies away.

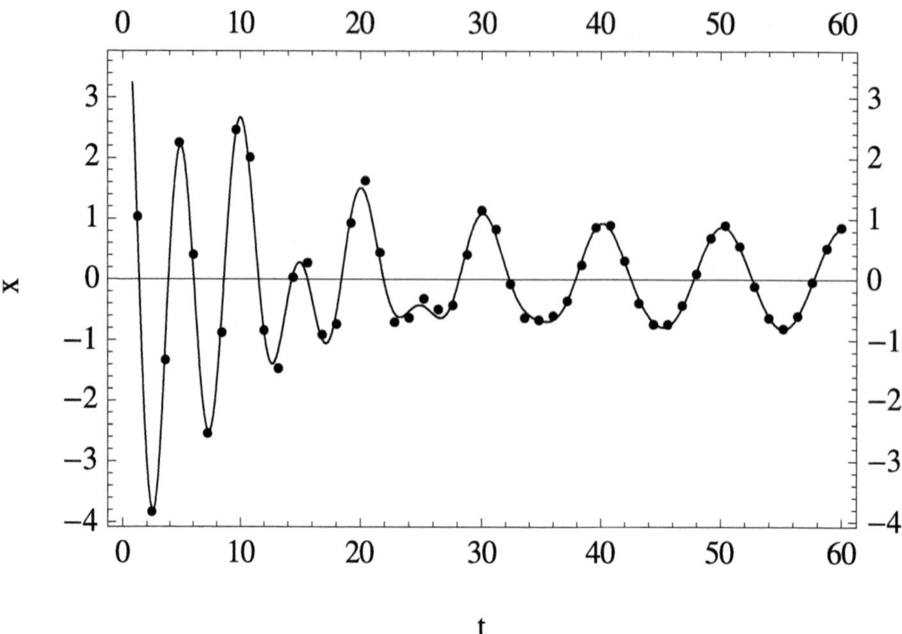

Fig. 1.30 Numerical solution of differential equation obeyed by driven damped harmonic oscillation obtained using Taylor series method using increment $h = 1.2$, showing displacement of the particle as function of time.. The curve is analytic solution

Program Number 1.7 (damped harmonic motion in phase space)

```
h=0.1;
i=0;
b=0.2;
V[0]=v=0;
wd=N[Sqrt[1-(b/2)^2]];
alpha1=ArcTan[-b/(2*wd)];
X[0]=x=Cos[alpha1]

Table[{
i=i+1,
t=t+h,

vd1=-x-b*v;
vd2=-v-b*vd1;
vd3=-vd1-b*vd2;
vd4=-vd2-b*vd3;
vd5=-vd3-b*vd4;
```

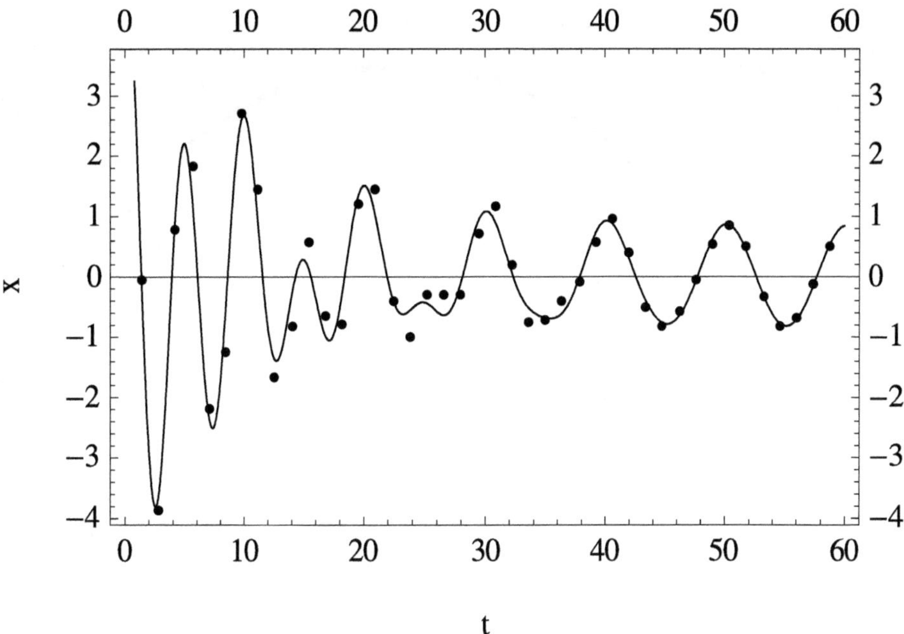

Fig. 1.31 Numerical solution of differential equation obeyed by driven damped harmonic oscillation obtained using Taylor series method using increment **h = 1.4**, showing displacement of the particle as function of time.. The curve is analytic solution

```
xd1=v;
xd2=vd1;
xd3=vd2;
xd4=vd3;
xd5=vd4;

V[i]=v=v+h*vd1+(1/2)*(h^2)*vd2+(1/6)*(h^3)*vd3+
(1/24)*(h^4)*vd4+(1/120)*(h^5)*vd5,

X[i]=x=x+h*xd1+(1/2)*(h^2)*xd2+(1/6)*(h^3)*xd3+
(1/24)*(h^4)*xd4+(1/120)*(h^5)*xd5},

{t,0,50-h,h}];
TableForm[%,TableSpacing->{2,2},
TableHeadings->{None,{"i","t","v","x"}}]

i=-1;
p1=ListLinePlot[Table[{i=i+1;t=t+h;X[i],V[i]},
{t,0-h,50-h,h}],
Frame->True,FrameLabel->{"x","v"},
FrameTicks->All,PlotStyle->{Black},
```

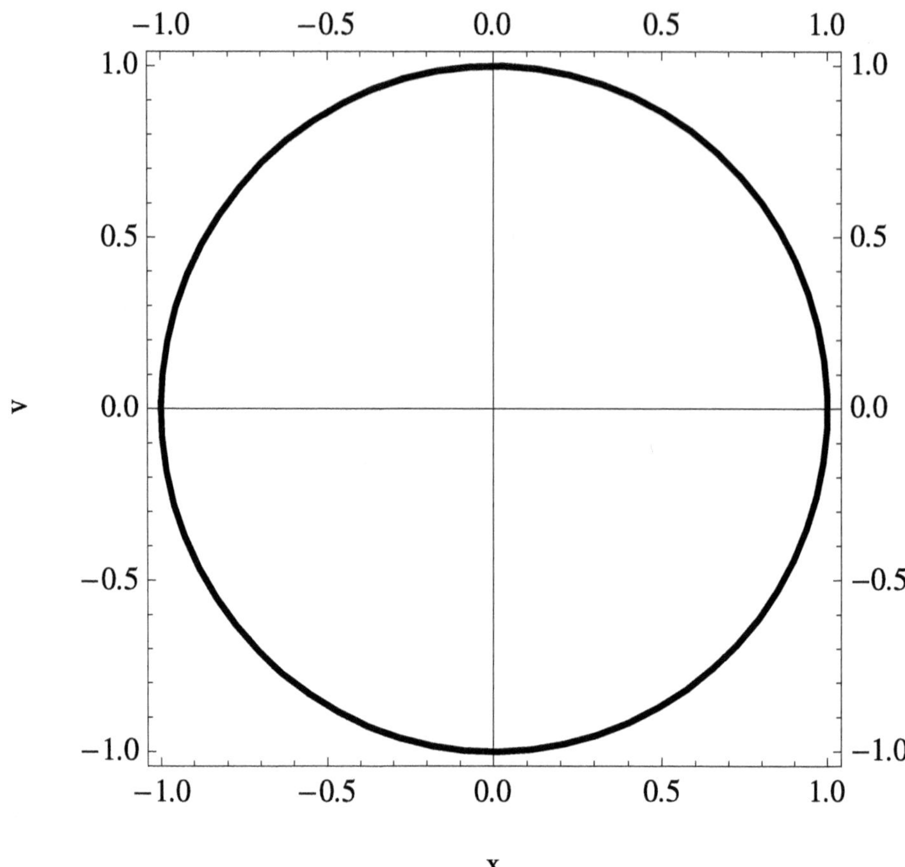

Fig. 1.32 Showing simple harmonic motion in phase space, i.e. speed v versus displacement x plot of a particle executing simple harmonic motion. Obtained using program number 1.6

```
PlotRange->{-1,1},AspectRatio->Automatic]
```

We now modify program number 1.5 to obtain speed v versus displacement x plot of a particle executing driven damped harmonic motion. The modified program is program number 1.8. The motion in phase space is as in Fig. 1.34. We find that oscillation wanes first until damped motion dies away. Thereafter, steady oscillation persists because of the steady oscillatory driving force.

Program Number 1.8 (driven damped harmonic motion in phase space)

```
T=5;
T0=5;

w=2*Pi/T;
```

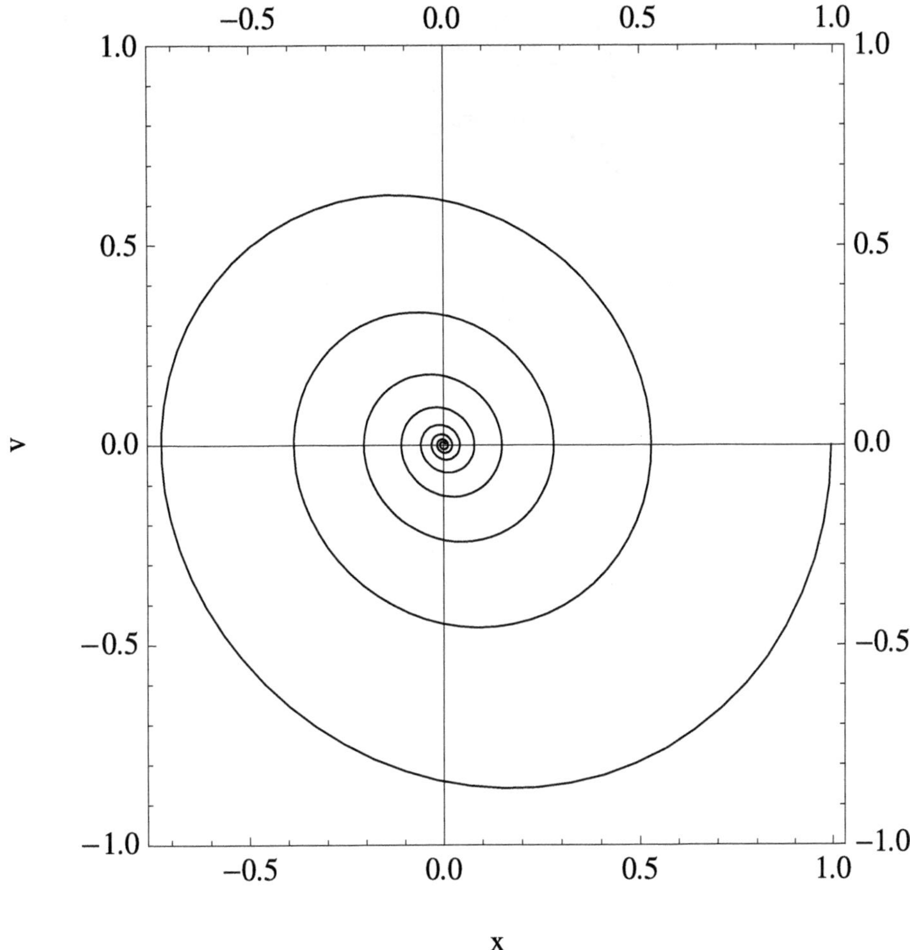

Fig. 1.33 Showing damped harmonic motion in phase space, i.e. speed v versus displacement x plot of a particle executing damped harmonic motion. Obtained using program number 1.7

```
w0=2*Pi/T0;

A=4;
a=1;
b=0.5;
h=0.1;

i=0;
wd=Sqrt[w0^2-(b/2)^2];
Td=2*Pi/wd
alpha1=0;
```

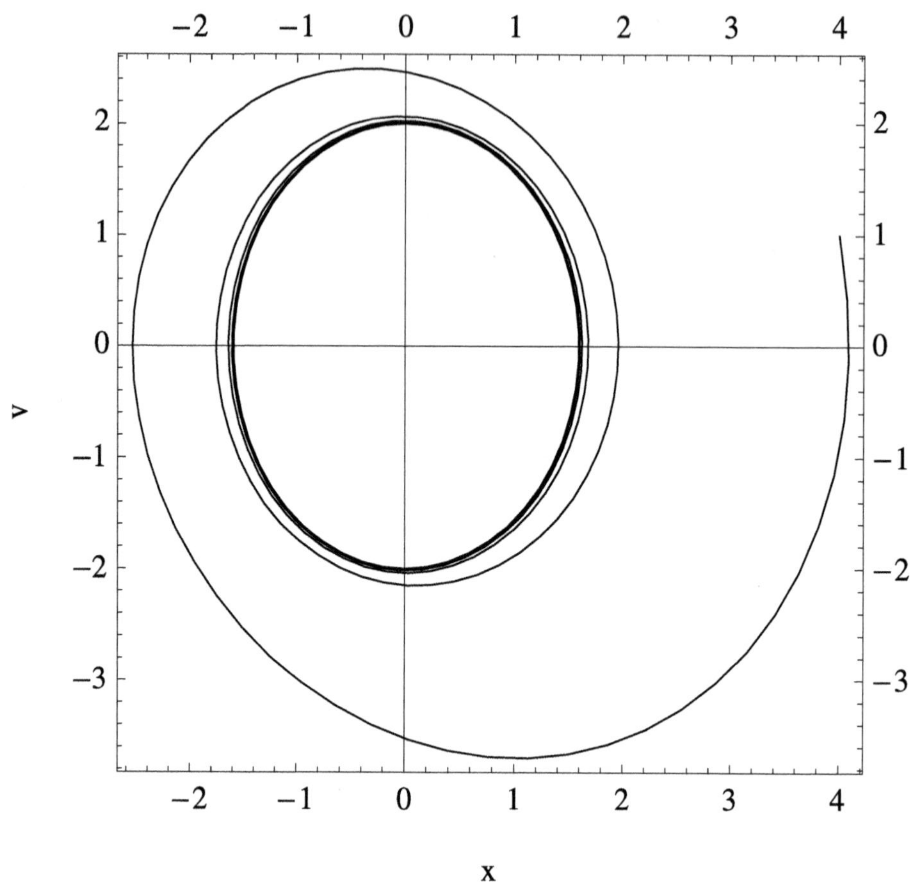

Fig. 1.34 Showing driven damped harmonic motion in phase space, i.e. speed v versus displacement x plot of a particle in driven damped oscillation. Obtained using program number 1.8

```
X[0]=x=A*Cos[alpha1]+(a/Sqrt[(w0^2-w^2)^2+
(b*w)^2])*Sin[-ArcTan[(w^2-w0^2)/(b*w)]]

V[0]=v=A*(-b/2)*Cos[alpha1]-A*wd*Sin[alpha1]+
(a/Sqrt[(w0^2-w^2)^2+
(b*w)^2])*w*Cos[-ArcTan[(w^2-w0^2)/(b*w)]]

Table[{
i=i+1,
t=t+h,

vd1=-b*v-(w0^2)*x+a*Cos[w*(t-h)];
vd2=-b*vd1-(w0^2)*v-w*a*Sin[w*(t-h)];
vd3=-b*vd2-(w0^2)*vd1-(w^2)*a*Cos[w*(t-h)];
```

```
vd4=-b*vd3-(w0^2)*vd2+(w^3)*a*Sin[w*(t-h)];
vd5=-b*vd4-(w0^2)*vd3+(w^4)*a*Cos[w*(t-h)];

xd1=v;
xd2=vd1;
xd3=vd2;
xd4=vd3;
xd5=vd4;

V[i]=v=v+h*vd1+(1/2)*h^2*vd2+(1/6)*h^3*vd3+
(1/24)*h^4*vd4+(1/120)*h^5*vd5,

X[i]=x=x+h*xd1+(1/2)*h^2*xd2+(1/6)*h^3*xd3+
(1/24)*h^4*xd4+(1/120)*h^5*xd5},

{t,0,50-h,h}];
TableForm[%,TableSpacing->{2,2},
TableHeadings->{None,{"i","t","v","x"}}]

i=-1;
p1=ListLinePlot[Table[{i=i+1;t=t+h;X[i],V[i]},
{t,0-h,50-h,h}],
Frame->True,FrameLabel->{"x","v"},
FrameTicks->All,PlotStyle->{Black},
AspectRatio->Automatic]
```

1.8   Numerical Simulation of Cyclotron Motion (of Electron) Using Taylor Series Method

An electron in uniform static magnetic field B along z direction executes uniform circular motion called cyclotron motion in xy plane with constant angular frequency $\omega_c = \dfrac{qB}{m}$ where q and m are charge and mass of an electron respectively. Random velocity of the electron in absence of magnetic field becomes speed of the electron with which it executes uniform circular motion in presence of the magnetic field. For typical material (GaAs), effective mass of electron is $0.067 m_e = 0.067 \times 9.1 \times 10^{-31}$ kg and the random velocity of electron is around 2.5×10^4 m/s. We have $q = -1.6 \times 10^{-19}$ Coulomb. See Fig. 1.35. Location of centre of the circular orbit is independent of time. Both radius R_c and period T of the orbit are inversely proportional to the magnetic field B. However, the magnetic field does not change speed of the electron in its circular orbit.

Lorentz force acting on an electron is $q\,\vec{v} \times \vec{B}$ and hence Newton's equation of motion is

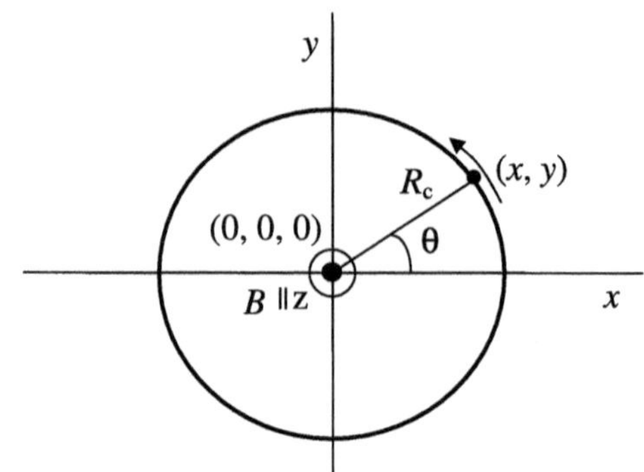

Fig. 1.35 Circular trajectory of an electron free in xy plane in a uniform static magnetic field B along z direction. (x, y) are coordinates of the electron. $\theta = \omega_c\, t$ where t is time

$$\vec{F} = m\frac{d\vec{v}}{dt} = q\vec{v} \times \vec{B} \tag{1.44}$$

Here

$$\vec{v} \times \vec{B} = \begin{vmatrix} \vec{e}_x & \vec{e}_y & \vec{e}_z \\ v_x & v_y & 0 \\ 0 & 0 & B \end{vmatrix} = \vec{e}_x v_y B - \vec{e}_y v_x B + \vec{e}_z 0 \tag{1.45}$$

considering that electron is restricted to xy plane and have no motion along z direction. As such

$$\frac{dv_x}{dt} = \frac{qB}{m} v_y = \omega_c v_y \tag{1.46}$$

$$\frac{dv_y}{dt} = \frac{qB}{m} v_x = -\omega_c v_x \tag{1.47}$$

where

$$\frac{dx}{dt} = v_x \tag{1.48}$$

and

$$\frac{dy}{dt} = v_y \tag{1.49}$$

Here we numerically solve differential equation (1.46) to (1.49) as initial value problems by Taylor series method and compare the results with known analytical solution:

$$x = R_c \cos\left(\omega_c t\right) \tag{1.50}$$

$$y = R_c \sin(\omega_c t) \tag{1.51}$$

Initial values that we have used are $(x, y) = (R_c, 0)$ and $(v_x, v_y) = (0, 2.5 \times 10^4)$ at time $t = 0$. 2.5×10^4 m/s is typical value of random velocity (Fermi velocity) of electron in GaAs-AlGaAs potential well or quantum well.

To solve Eq. (1.46) using Taylor series method, iteration for numerical solution is given by

$$vx_{new} = vx_{old} + h\ vxd1 + (1/2)h^2 vxd2 + (1/6)h^3 vxd3$$
$$+ (1/24)h^4 vxd4 + (1/120)h^5 vxd5 \tag{1.52}$$

where we have denoted $\dfrac{dv_x}{dt}$ by $vxd1$, $\dfrac{d^2 v_x}{dt^2}$ by $vxd2$, $\dfrac{d^3 v_x}{dt^3}$ by $vxd3$, $\dfrac{d^4 v_x}{dt^4}$ by $vxd4$ and $\dfrac{d^5 v_x}{dt^5}$ by $vxd5$.

To solve Eq. (1.47) using Taylor series method, iteration for numerical solution is given by

$$vy_{new} = vy_{old} + h\ vyd1 + (1/2)h^2 vyd2 + (1/6)h^3 vyd3 + (1/24)h^4 vyd4$$
$$+ (1/120)h^5 vyd5 \tag{1.53}$$

where we have denoted $\dfrac{dv_y}{dt}$ by $vyd1$, $\dfrac{d^2 v_y}{dt^2}$ by $vyd2$, $\dfrac{d^3 v_y}{dt^3}$ by $vyd3$, $\dfrac{d^4 v_y}{dt^4}$ by $vyd4$ and $\dfrac{d^5 v_y}{dt^5}$ by $vyd5$.

To solve Eq. (1.48) using Taylor series method, iteration for numerical solution is given by

$$x_{new} = x_{old} + h\ xd1 + (1/2)h^2 xd2 + (1/6)h^3 xd3 + (1/24)h^4 xd4 + (1/120)h^5 xd5 \tag{1.54}$$

where we have denoted $\dfrac{dx}{dt}$ by $xd1$, $\dfrac{d^2 x}{dt^2}$ by $xd2$, $\dfrac{d^3 x}{dt^3}$ by $xd3$, $\dfrac{d^4 x}{dt^4}$ by $xd4$ and $\dfrac{d^5 x}{dt^5}$ by $xd5$.

To solve Eq. (1.49) using Taylor series method, iteration for numerical solution is given by

$$y_{new} = y_{old} + h\ yd1 + (1/2)h^2 yd2 + (1/6)h^3 yd3 + (1/24)h^4 yd4 + (1/120)h^5 yd5 \tag{1.55}$$

where we have denoted $\dfrac{dy}{dt}$ by $yd1$, $\dfrac{d^2 y}{dt^2}$ by $yd2$, $\dfrac{d^3 y}{dt^3}$ by $yd3$, $\dfrac{d^4 y}{dt^4}$ by $yd4$ and $\dfrac{d^5 y}{dt^5}$ by $yd5$.

We shall explore the problem in the time interval $0 \le t \le$ To where To is time of observation, typically 5 ps. We shall use initial values as already mentioned. The program becomes as in program number 1.9. This is symbolic computation and is evident. Resulting data are in Table 1.6 and in Figs. 1.36, 1.37, 1.38, 1.39, 1.40, 1.41, and 1.42.

Table 1.6 Numerical data for location (x, y) of electron in its trajectory obtained using Taylor series method using increment $h = 5/20$ ps using program number 1.9. For parameters as in program 1.9

i	t (ps)	x (nm)	y (nm)
0	0.00	19.053	0.000
1	0.25	18.037	6.139
2	0.50	15.098	11.622
3	0.75	10.548	15.867
4	1.00	4.874	18.419
5	1.25	−1.320	19.008
6	1.50	−7.374	17.569
7	1.75	−12.641	14.256
8	2.00	−16.560	9.423
9	2.25	−18.713	3.586
10	2.50	−18.871	−2.634
11	2.75	−17.016	−8.574
12	3.00	−13.346	−13.599
13	3.25	−8.253	−17.173
14	3.50	−2.280	−18.917
15	3.75	3.936	−18.643
16	4.00	9.732	−16.381
17	4.25	14.491	−12.372
18	4.50	17.704	−7.044
19	4.75	19.029	−0.964
20	5.00	18.325	5.218

Program Number 1.9 (cyclotron motion)

```
To=5.0*10^-12;
h=To/20;

m=0.067*9.1*10^-31;
q=1.6*10^-19;
B=0.5;
wc=q*B/m;

vx=0;
vy=2.5*10^4;

Rc=(Sqrt[vx^2+vy^2])/wc
X[0]=x=Rc;
Y[0]=y=0;

i=0;
```

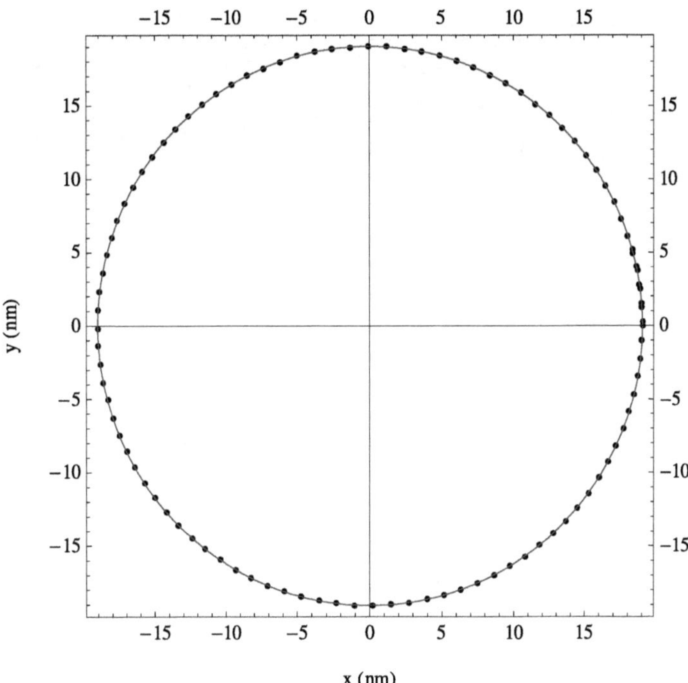

Fig. 1.36 Numerical simulation of cyclotron motion of electron showing location (x, y) of electron in its trajectory obtained using Taylor series method using increment $h = 5/100$ ps using program number 1.9. For parameters as in program number 1.9

```
Table[{
i=i+1,
t=t+h,

vxd1=-wc*vy;          vyd1=wc*vx;
vxd2=-wc*vyd1;   vyd2=wc*vxd1;
vxd3=-wc*vyd2;   vyd3=wc*vxd2;
vxd4=-wc*vyd3;   vyd4=wc*vxd3;
vxd5=-wc*vyd4;   vyd5=wc*vxd4;

xd1=vx;
xd2=vxd1;
xd3=vxd2;
xd4=vxd3;
xd5=vxd4;

yd1=vy;
yd2=vyd1;
yd3=vyd2;
yd4=vyd3;
yd5=vyd4;
```

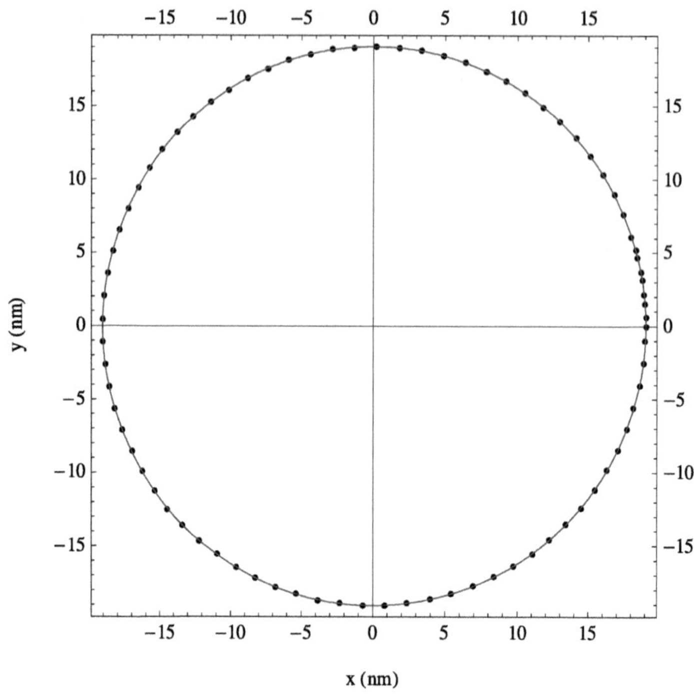

Fig. 1.37 Numerical simulation of cyclotron motion of electron showing location (x, y) of electron in its trajectory obtained using Taylor series method using increment $h = 5/80$ ps using program number 1.9. For parameters as in program number 1.9

```
Vx[i]=vx=vx+h*vxd1+(1/2)*h^2*vxd2+
(1/6)*h^3*vxd3+(1/24)*h^4*vxd4+(1/120)*h^5*vxd5;

X[i]=x=x+h*xd1+(1/2)*h^2*xd2+(1/6)*h^3*xd3+
(1/24)*h^4*xd4+(1/120)*h^5*xd5,

Vy[i]=vy=vy+h*vyd1+(1/2)*h^2*vyd2+
(1/6)*h^3*vyd3+(1/24)*h^4*vyd4+(1/120)*h^5*vyd5;

Y[i]=y=y+h*yd1+(1/2)*h^2*yd2+(1/6)*h^3*yd3+
(1/24)*h^4*yd4+(1/120)*h^5*yd5},

{t,0,To-h,h}];
TableForm[%,TableSpacing->{2,2},
TableHeadings->{None,{"i","t (sec)","x","y"}}];

i=-1;
Table[{i=i+1,t=t+h;t*10^12,X[i]*10^9,Y[i]*10^9},
{t,0-h,To-h,h}];
TableForm[%,TableSpacing->{2,2},
```

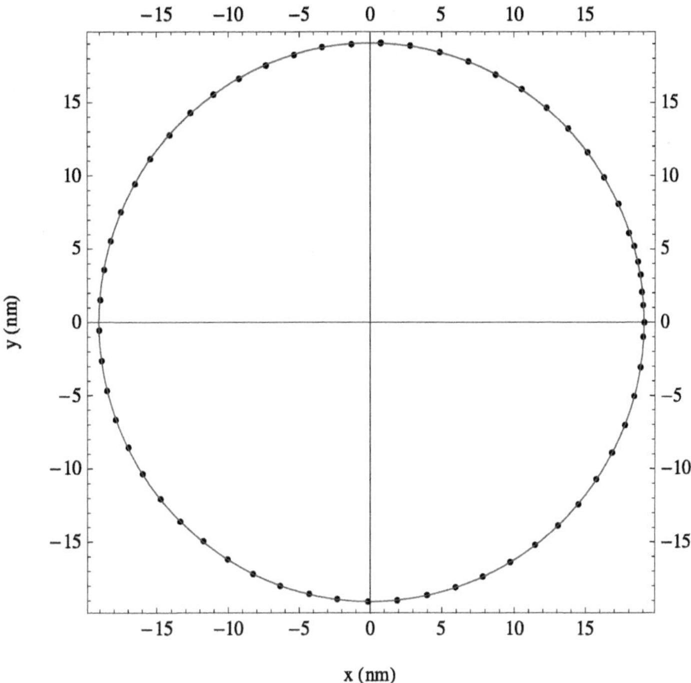

Fig. 1.38 Numerical simulation of cyclotron motion of electron showing location (x, y) of electron in its trajectory obtained using Taylor series method using increment $h = 5/60$ ps using program number 1.9. For parameters as in program number 1.9

```
TableHeadings->{None,{"i","t (ps)","x (nm)","y (nm)"}}]

i=-1;
p1=ListPlot[Table[{i=i+1;t=t+h;X[i]*10^9,Y[i]*10^9},
{t,0-h,To-h,h}],
Frame->True,FrameLabel->{"x (nm)","y (nm)"},
FrameTicks->All,PlotStyle->{Black},AspectRatio->Automatic];

p2=ParametricPlot[{(Rc*Cos[wc*t])*10^9,(Rc*Sin[wc*t])*10^9},
{t,0,To},AspectRatio->Automatic];
Show[p1,p2]
```

We have numerically solved Newton's second law of motion for Lorentz force exerted on an electron by a uniform static magnetic field. We have used Taylor series method for solving the problem as initial value problem. We could obtain trajectory of electron in cyclotron motion. We find that we can use large values of increment of the independent variable (time) and get numerical solutions that closely match with known exact analytic solution shown by circles in Figs. 1.36, 1.37, 1.38, 1.39, 1.40, 1.41, and 1.42. We have performed symbolic computation in Mathematica® using which we have got the trajectory of electron.

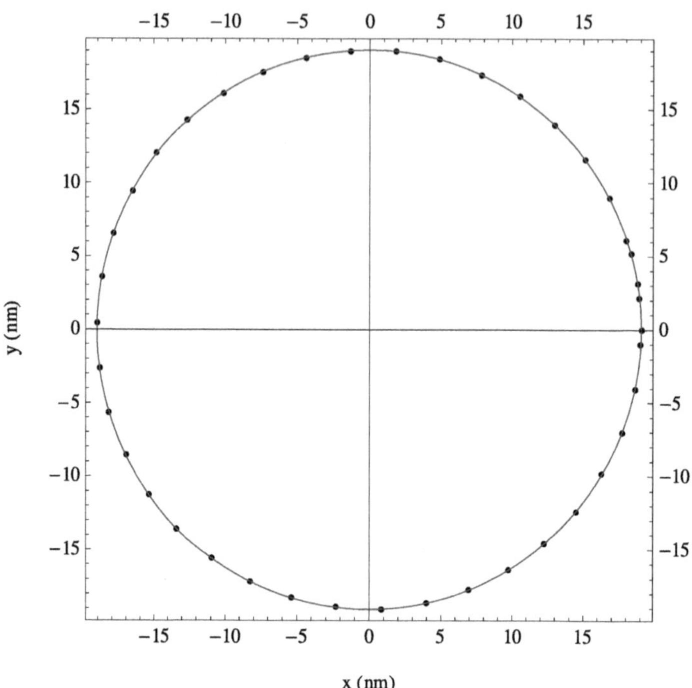

Fig. 1.39 Numerical simulation of cyclotron motion of electron showing location (x, y) of electron in its trajectory obtained using Taylor series method using increment $h = 5/40$ ps using program number 1.9. For parameters as in program number 1.9

1.9   Numerical Solution of Differential Equation for Hyperbolic Function cosh Using Taylor Series Method

Differential equation obeyed by the Hyperbolic function

$$y = \cosh(x) = \frac{e^x + e^{-x}}{2} \tag{1.56}$$

is

$$\frac{d^2 y}{dx^2} = y \tag{1.57}$$

Here we have numerically solved differential equation (1.57) as an initial value problem by Taylor series method and compared the results with analytical solution given by Eq. (1.56).

But Eq. (1.57) is a second order differential equation whereas we need first order differential equation to deal with the Taylor series. Therefore we rewrite Eq. (1.57) as two first order differential equations as

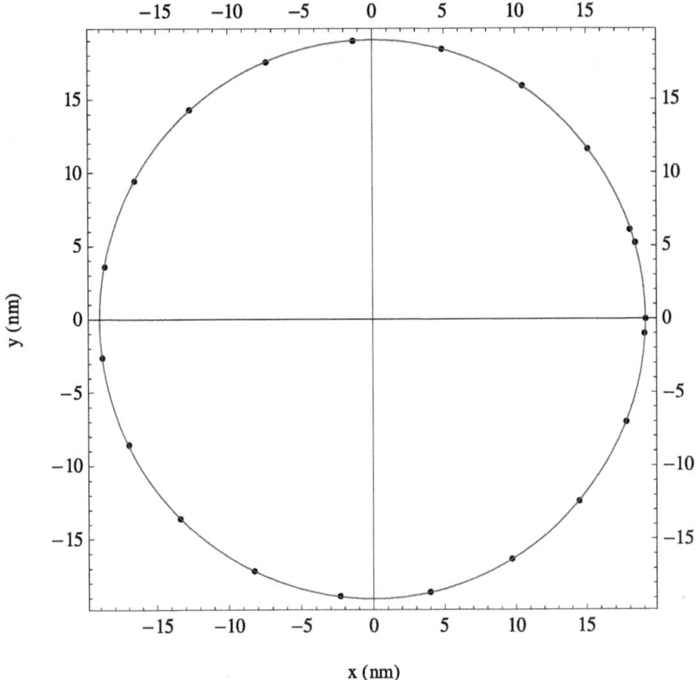

Fig. 1.40 Numerical simulation of cyclotron motion of electron showing location (x, y) of electron in its trajectory obtained using Taylor series method using increment $h = 5/20$ ps using program number 1.9. For parameters as in program number 1.9

$$\frac{dv}{dx} = y \tag{1.58}$$

and

$$\frac{dy}{dx} = v \tag{1.59}$$

To solve Eq. (1.58) using Taylor series method, iteration for numerical solution is given by

$$v_{new} = v_{old} + h\,vd1 + (1/2)h^2 vd2 + (1/6)h^3 vd3 + (1/24)\,h^4 vd4 + (1/120)h^5 vd5 \tag{1.60}$$

where we have denoted $\dfrac{dv}{dx}$ by $vd1$, $\dfrac{d^2 v}{dx^2}$ by $vd2$, $\dfrac{d^3 v}{dx^3}$ by $vd3$, $\dfrac{d^4 v}{dx^4}$ by $vd4$ and $\dfrac{d^5 v}{dx^5}$ by $vd5$.

To solve Eq. (1.59) using Taylor series method, iteration for numerical solution is given by

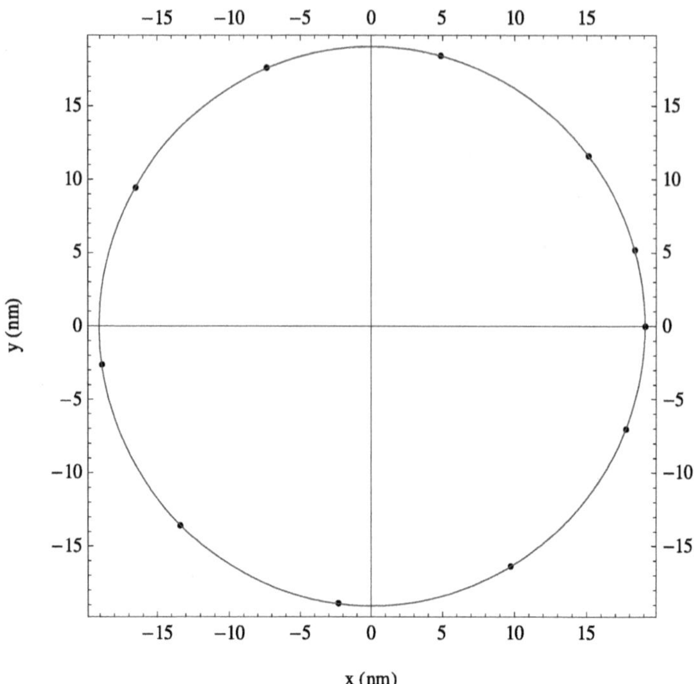

Fig. 1.41 Numerical simulation of cyclotron motion of electron showing location (x, y) of electron in its trajectory obtained using Taylor series method using increment $h = 5/10$ ps using program number 1.9. For parameters as in program number 1.9

$$y_{\text{new}} = y_{\text{old}} + h\, yd1 + (1/2)h^2 yd2 + (1/6)h^3 yd3 + (1/24)h^4 yd4 + (1/120)h^5 yd5 \qquad (1.61)$$

where we have denoted $\dfrac{dy}{dx}$ by $yd1$, $\dfrac{d^2 y}{dx^2}$ by $yd2$, $\dfrac{d^3 y}{dx^3}$ by $yd3$, $\dfrac{d^4 y}{dx^4}$ by $yd4$ and $\dfrac{d^5 y}{dx^5}$ by $yd5$.

We shall explore the Hyperbolic function $\cosh(x)$ in the interval $-3 \le x \le +3$ for which the known initial values are $(x, v) = (-3, -10.0179)$ and $(x, y) = (-3, 10.0677)$. We shall also explore the Hyperbolic function $\cosh(x)$ in the interval $-5 \le x \le +5$ for which the known initial values are $(x, v) = (-5, -74.2032)$ and $(x, y) = (-5, 74.2099)$.

The programs become as in program number 1.10 and program number 1.11 depending on the choice of the interval and hence of the initial values. This is symbolic computation and is evident. Resulting data are in Tables 1.7 and 1.8 and in Figs. 1.43, 1.44, 1.45, 1.46, 1.47, 1.48, 1.49, 1.50, 1.51, 1.52, 1.53, 1.54, 1.55, and 1.56.

Program Number 1.10 (Hyperbolic function cosh for $-3 \le x \le +3$)

```
h=0.25;
x=-3;
Y[0]=y=10.0677;
```

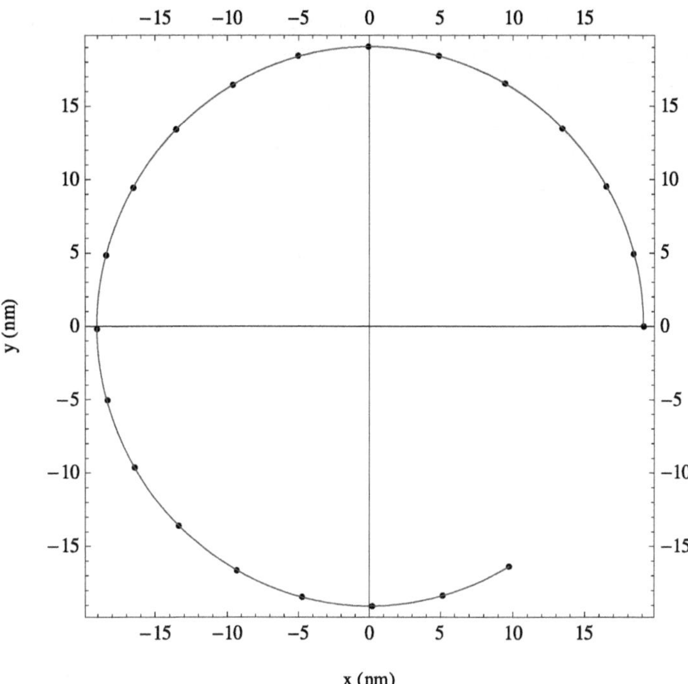

Fig. 1.42 Showing that for time of observation To = 4 ps, we find that electron cannot manage to complete one cycle of circular orbit. The plot has been obtained using Taylor series method using increment $h = 4/20$ ps using program number 1.9

```
V[0]=v=-10.0179;
i=0;

Table[{
i=i+1,
x=x+h,

vd1=y;
vd2=v;
vd3=y;
vd4=v;
vd5=y;

yd1=v;
yd2=vd1;
yd3=vd2;
yd4=vd3;
yd5=vd4;

v=v+h*vd1+(1/2)*(h^2)*vd2+(1/6)*(h^3)*vd3+
(1/24)*(h^4)*vd4+(1/120)*(h^5)*vd5,
```

Table 1.7 Numerical solution of differential equation obeyed by cosh(x) obtained using Taylor series method using increment $h = 0.25$ using program number 1.10 For initial values $(x, v) = (-3, -10.0179)$ and $(x, y) = (-3, 10.0677)$

i	x	v	y
1	−2.75	−7.789	7.853
2	−2.50	−6.050	6.132
3	−2.25	−4.691	4.797
4	−2.00	−3.627	3.762
5	−1.75	−2.790	2.964
6	−1.50	−2.129	2.352
7	−1.25	−1.602	1.888
8	−1.00	−1.175	1.543
9	−0.75	−0.822	1.295
10	−0.50	−0.521	1.128
11	−0.25	−0.253	1.032
12	0.00	0.000	1.000
13	0.25	0.253	1.032
14	0.50	0.521	1.128
15	0.75	0.823	1.295
16	1.00	1.176	1.543
17	1.25	1.602	1.889
18	1.50	2.130	2.353
19	1.75	2.791	2.965
20	2.00	3.628	3.763
21	2.25	4.692	4.798
22	2.50	6.052	6.134
23	2.75	7.791	7.855
24	3.00	10.020	10.070

```
Y[i]=y=y+h*yd1+(1/2)*(h^2)*yd2+(1/6)*(h^3)*yd3+
(1/24)*(h^4)*yd4+(1/120)*(h^5)*yd5},{x,-3,3-h,h}];


TableForm[%,TableSpacing->{2,2},
TableHeadings->{None,{"i","x","v","y"}}]


i=-1;
p1=ListPlot[Table[{i=i+1;x=x+h,Y[i]},{x,-3-h,3-h,h}],
Frame->True,FrameLabel->{"x","y"},FrameTicks->All,
PlotStyle->{Black},PlotRange->{0,11}];


p2=Plot[{Cosh[x]},{x,-3,3},PlotStyle->{Black},
PlotRange->{0,11}];


Show[p1,p2]
```

Table 1.8 Numerical solution of differential equation obeyed by cosh(x) obtained using Taylor series method using increment $h = 0.25$ using program number 1.11. For initial values $(x, v) = (-5, -74.2032)$ and $(x, y) = (-5, 74.2099)$

s	x	v	y
1	−4.75	−57.788	57.796
2	−4.50	−45.003	45.014
3	−4.25	−35.046	35.060
4	−4.00	−27.290	27.308
5	−3.75	−21.249	21.272
6	−3.50	−16.543	16.573
7	−3.25	−12.876	12.914
8	−3.00	−10.018	10.068
9	−2.75	−7.790	7.853
10	−2.50	−6.050	6.132
11	−2.25	−4.691	4.796
12	−2.00	−3.627	3.762
13	−1.75	−2.791	2.964
14	−1.50	−2.130	2.352
15	−1.25	−1.603	1.888
16	−1.00	−1.176	1.542
17	−0.75	−0.824	1.293
18	−0.50	−0.523	1.126
19	−0.25	−0.255	1.029
20	0.00	−0.003	0.997
21	0.25	0.249	1.028
22	0.50	0.516	1.123
23	0.75	0.816	1.289
24	1.00	1.168	1.535
25	1.25	1.592	1.879
26	1.50	2.117	2.340
27	1.75	2.774	2.948
28	2.00	3.606	3.741
29	2.25	4.664	4.770
30	2.50	6.016	6.098
31	2.75	7.745	7.809
32	3.00	9.961	10.011
33	3.25	12.803	12.842
34	3.50	16.449	16.479
35	3.75	21.129	21.152
36	4.00	27.136	27.154
37	4.25	34.848	34.862
38	4.50	44.749	44.760
39	4.75	57.462	57.470
40	5.00	73.785	73.791

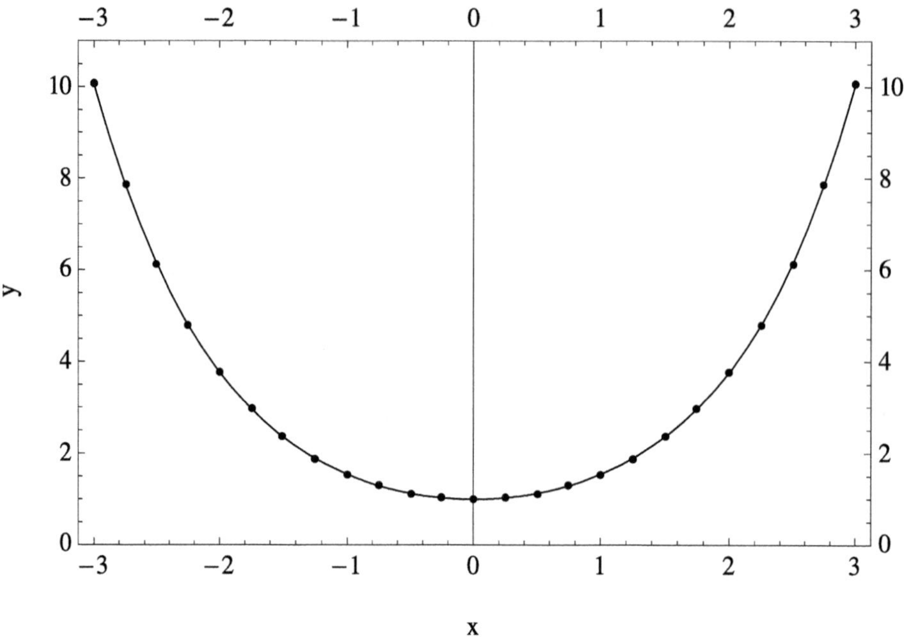

Fig. 1.43 Numerical solution of differential equation obeyed by cosh(x) obtained using Taylor series method using increment $h = 0.25$. The curve shows analytic solution

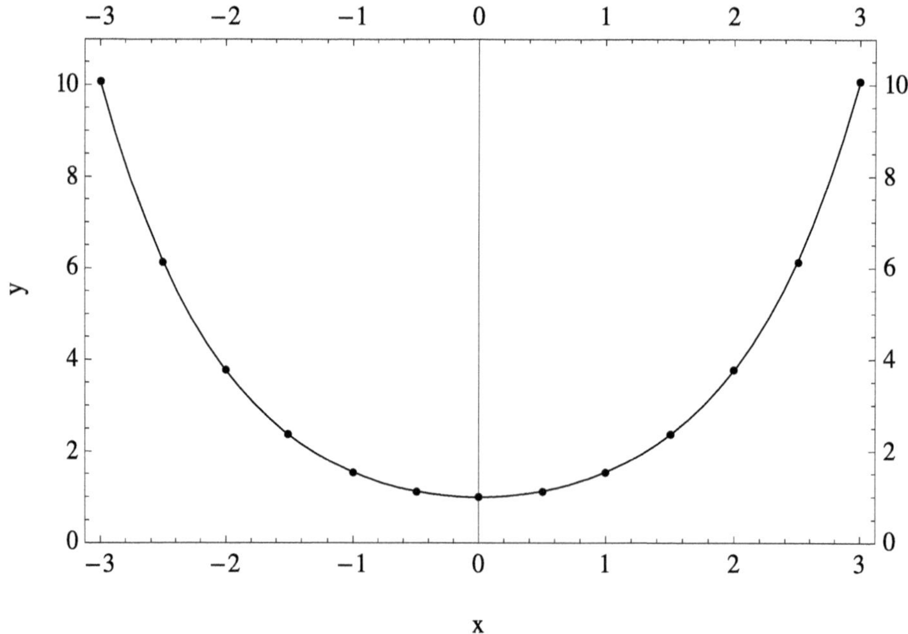

Fig. 1.44 Numerical solution of differential equation obeyed by cosh(x) obtained using Taylor series method using increment $h = 0.5$. The curve shows analytic solution

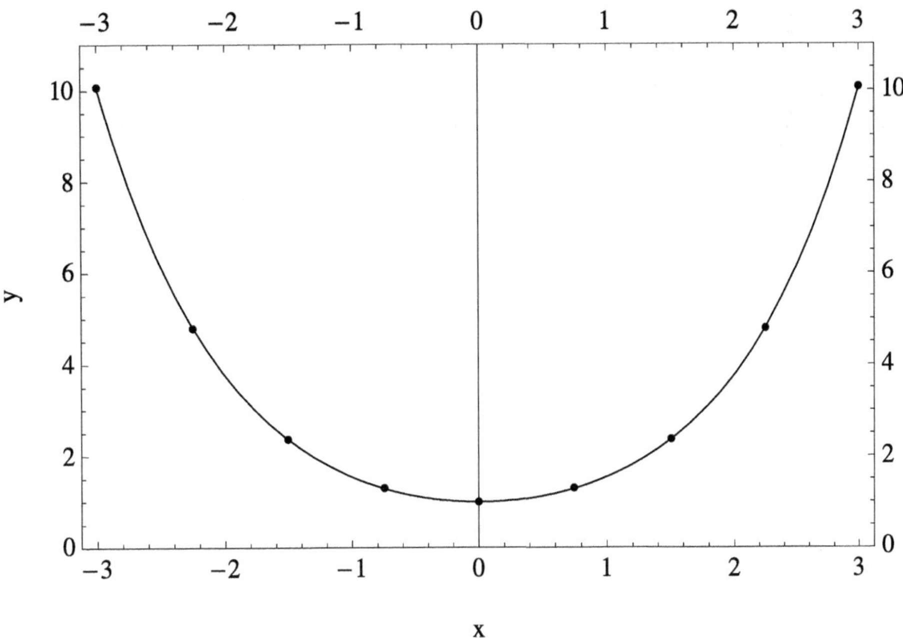

Fig. 1.45 Numerical solution of differential equation obeyed by cosh(x) obtained using Taylor series method using increment $h = 0.75$. The curve shows analytic solution

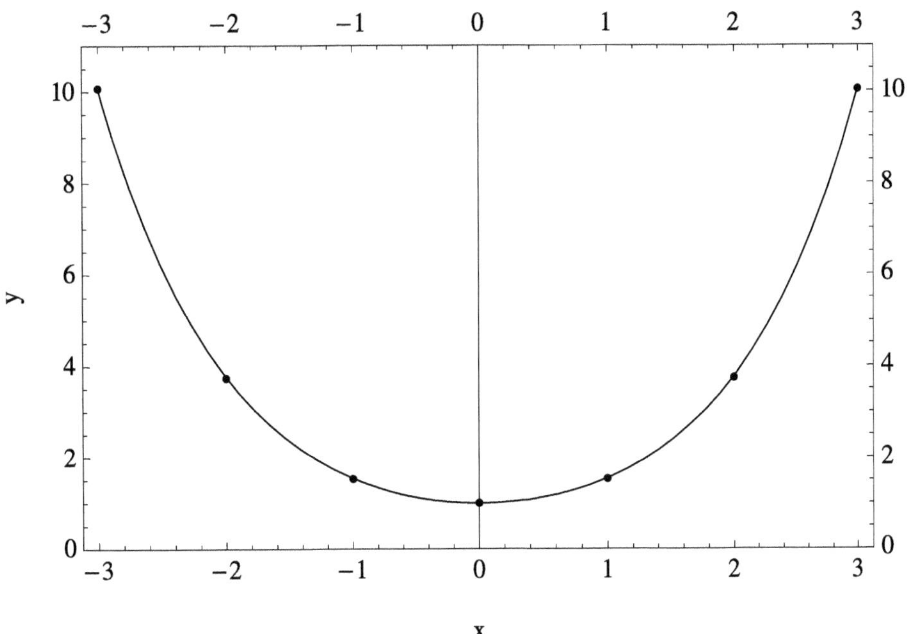

Fig. 1.46 Numerical solution of differential equation obeyed by cosh(x) obtained using Taylor series method using increment $h = 1.00$. The curve shows analytic solution

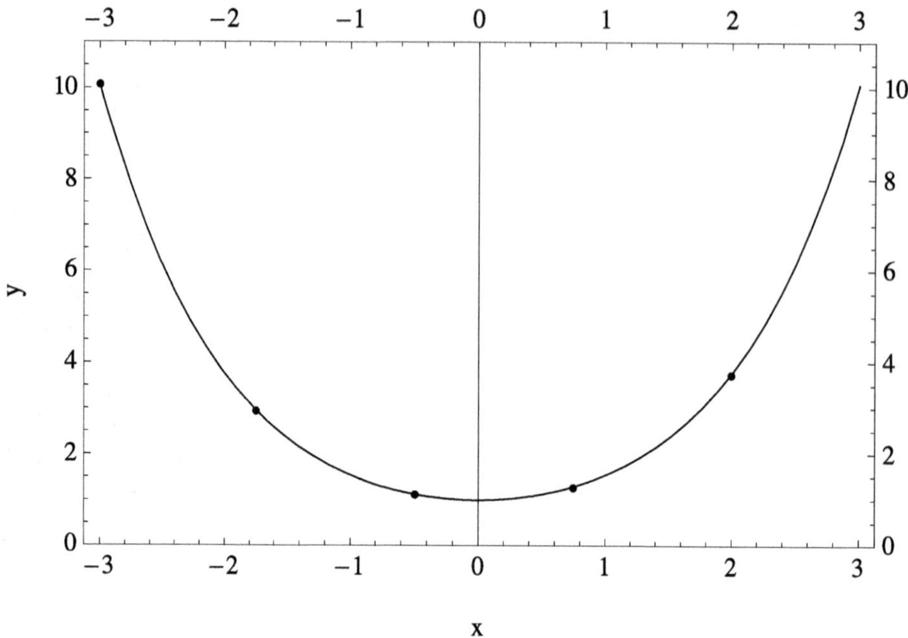

Fig. 1.47 Numerical solution of differential equation obeyed by cosh(x) obtained using Taylor series method using increment $h = 1.25$. The curve shows analytic solution

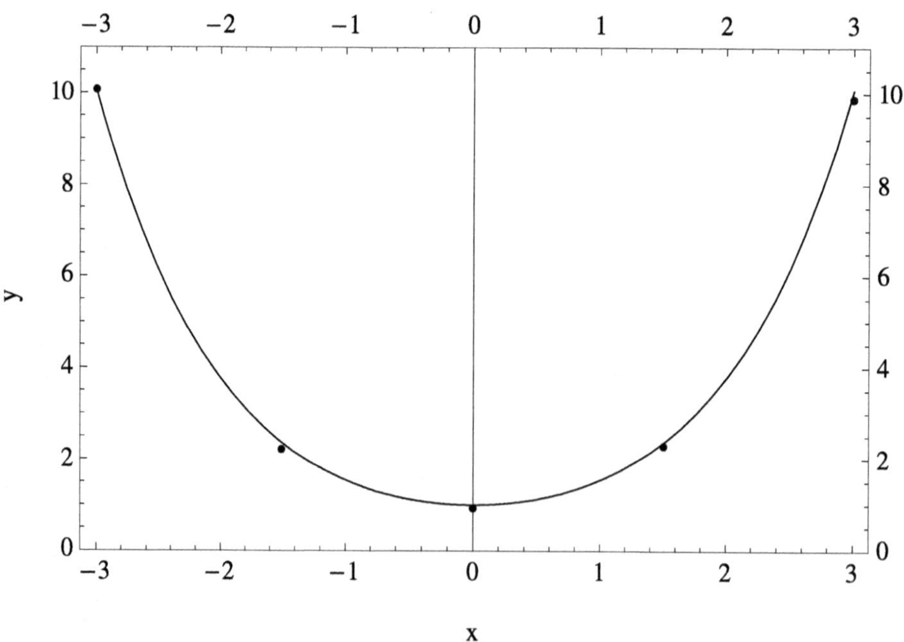

Fig. 1.48 Numerical solution of differential equation obeyed by cosh(x) obtained using Taylor series method using increment $h = 1.5$. The curve shows analytic solution

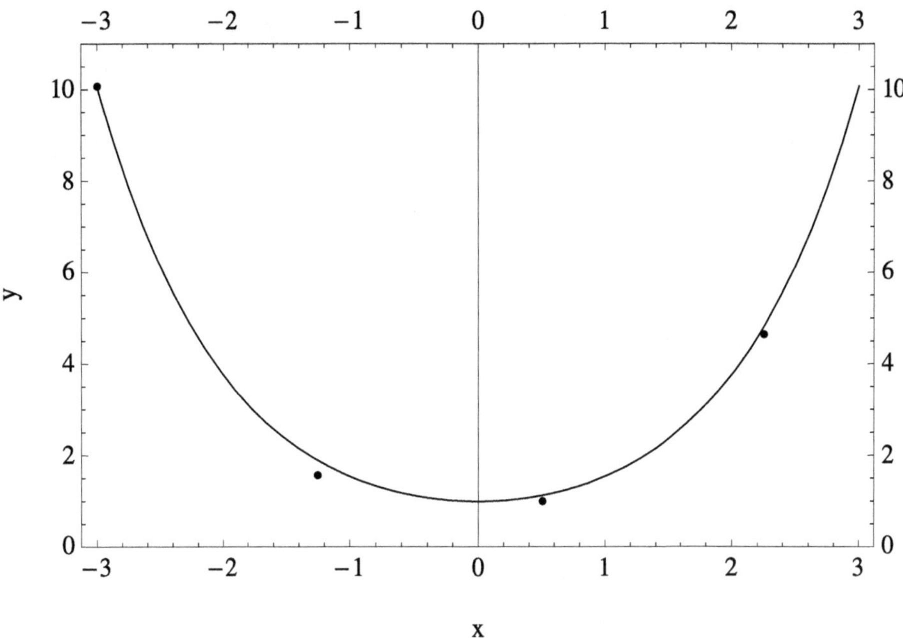

Fig. 1.49 Numerical solution of differential equation obeyed by cosh(x) obtained using Taylor series method using increment $h = 1.75$. The curve shows analytic solution

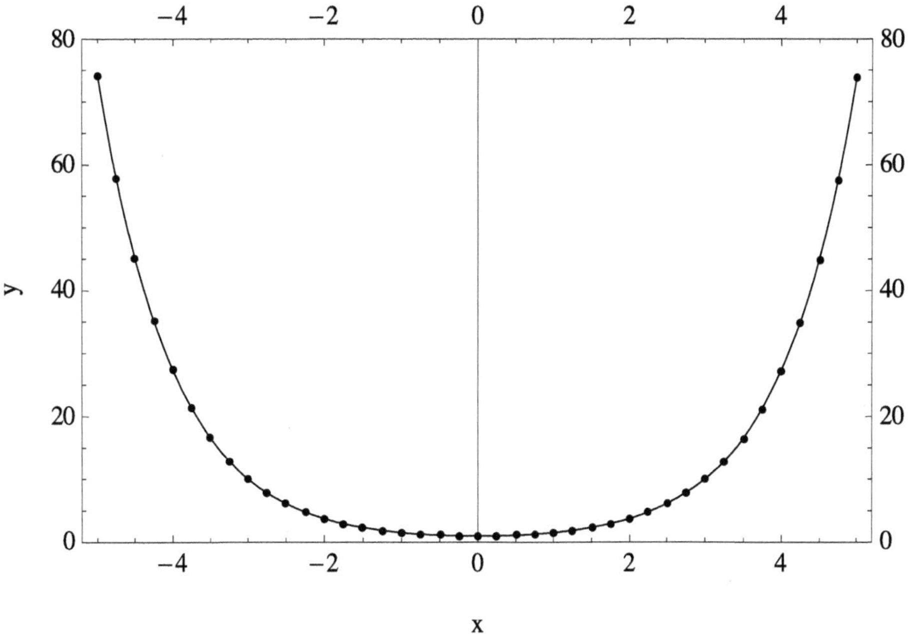

Fig. 1.50 Numerical solution of differential equation obeyed by cosh(x) obtained using Taylor series method using increment $h = 0.25$. The curve is analytic solution

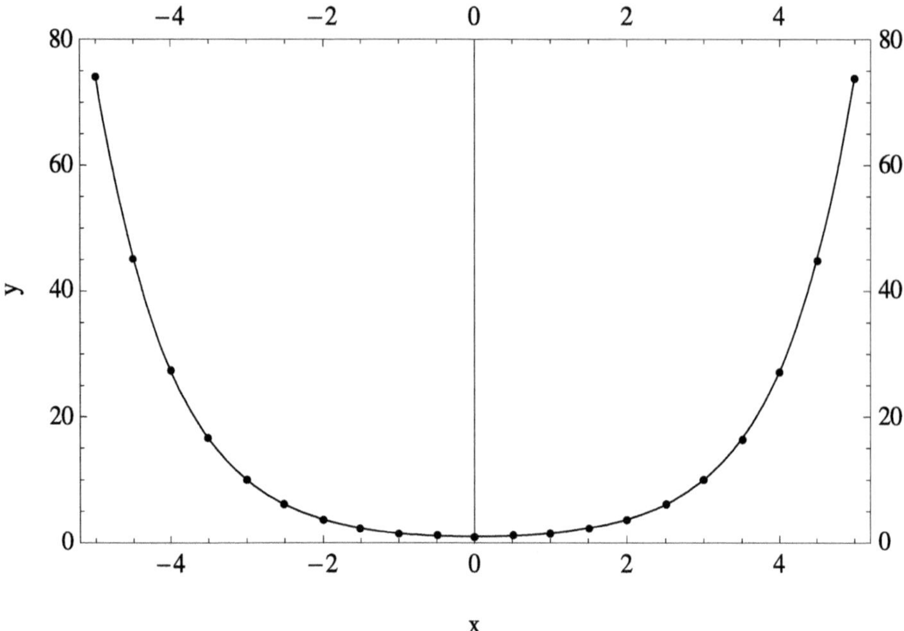

Fig. 1.51 Numerical solution of differential equation obeyed by cosh(x) obtained using Taylor series method using increment $h = 0.5$. The curve is analytic solution

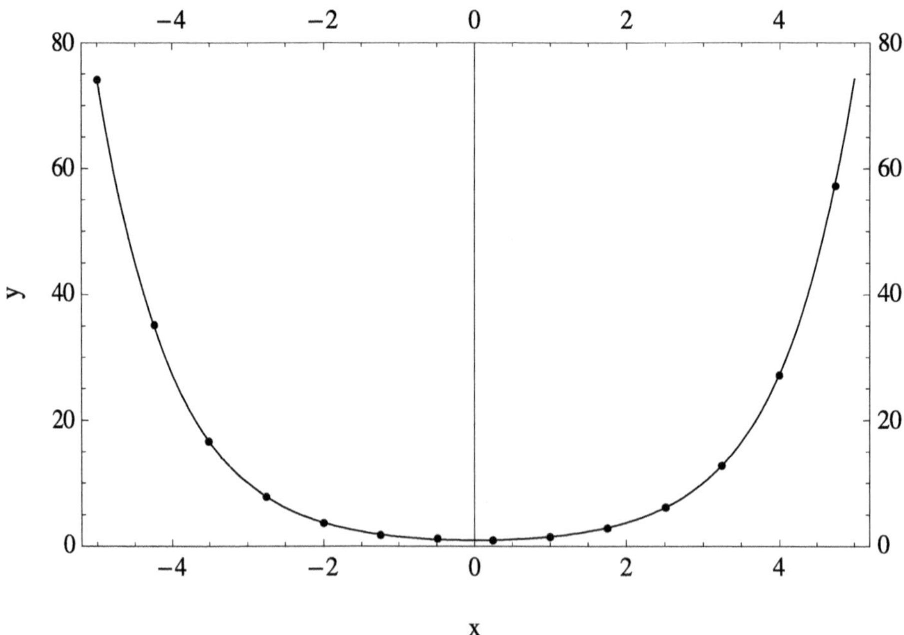

Fig. 1.52 Numerical solution of differential equation obeyed by cosh(x) obtained using Taylor series method using increment $h = 0.75$. The curve is analytic solution

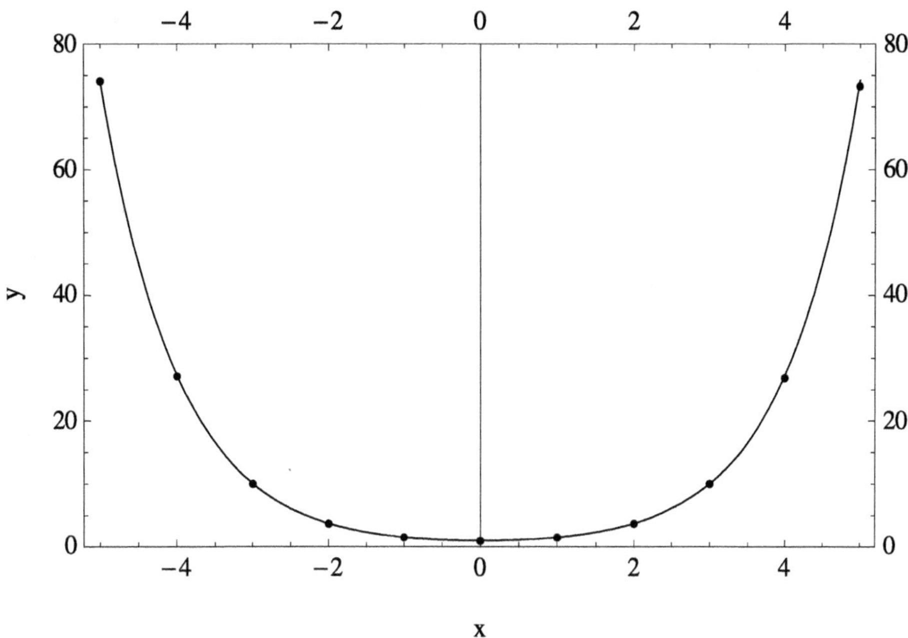

Fig. 1.53 Numerical solution of differential equation obeyed by cosh(x) obtained using Taylor series method using increment $h = 1$. The curve is analytic solution

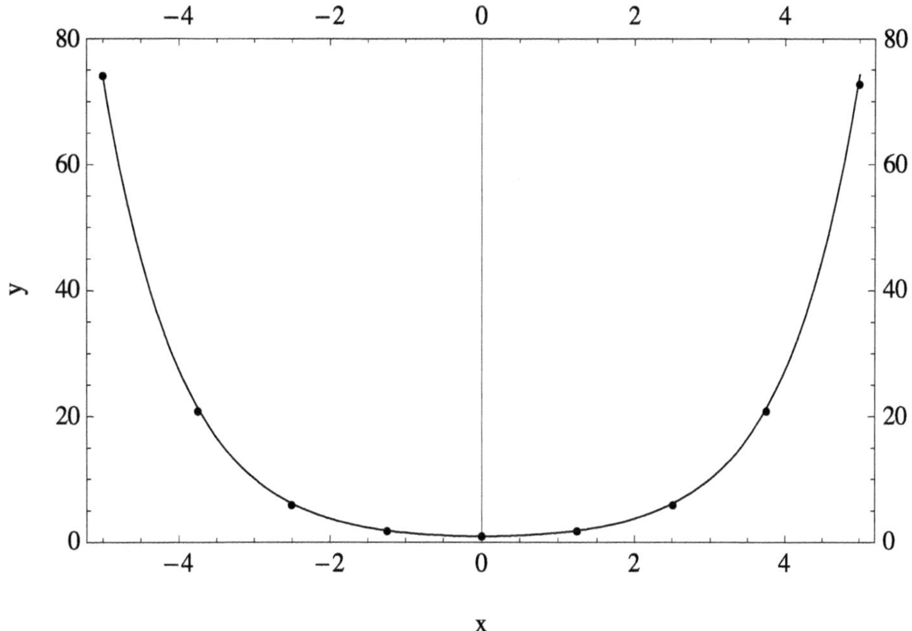

Fig. 1.54 Numerical solution of differential equation obeyed by cosh(x) obtained using Taylor series method using increment $h = 1.25$. The curve is analytic solution

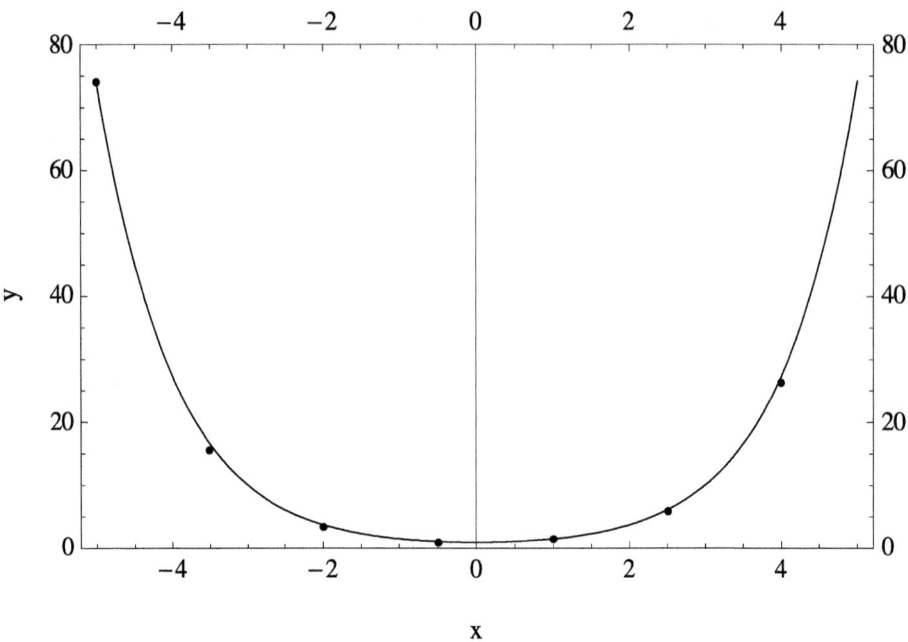

Fig. 1.55 Numerical solution of differential equation obeyed by cosh(x) obtained using Taylor series method using increment $h = 1.5$. The curve is analytic solution

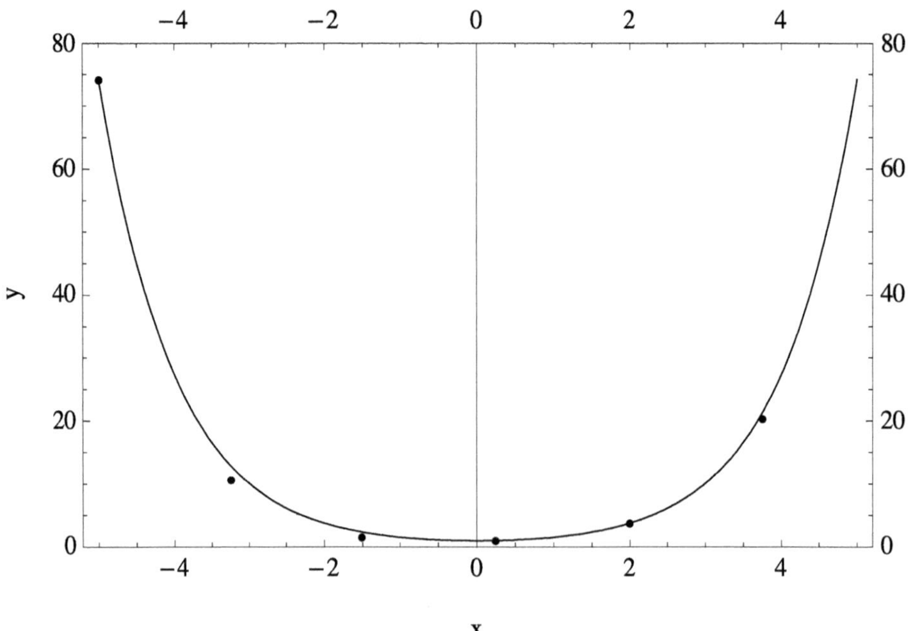

Fig. 1.56 Numerical solution of differential equation obeyed by cosh(x) obtained using Taylor series method using increment $h = 1.75$. The curve is analytic solution

Program Number 1.11 (Hyperbolic function cosh for $-5 \leq x \leq +5$)

```
h=0.25;
x=-5;
Y[0]=y=74.2099;
V[0]=v=-74.2032;
i=0;

Table[{
i=i+1,
x=x+h,

vd1=y;
vd2=v;
vd3=y;
vd4=v;
vd5=y;

yd1=v;
yd2=vd1;
yd3=vd2;
yd4=vd3;
yd5=vd4;

v=v+h*vd1+(1/2)*(h^2)*vd2+(1/6)*(h^3)*vd3+
(1/24)*(h^4)*vd4+(1/120)*(h^5)*vd5,

Y[i]=y=y+h*yd1+(1/2)*(h^2)*yd2+(1/6)*(h^3)*yd3+
(1/24)*(h^4)*yd4+(1/120)*(h^5)*yd5},{x,-5,5-h,h}];

TableForm[%,TableSpacing->{2,2},
TableHeadings->{None,{"i","x","v","y"}}]

i=-1;
p1=ListPlot[Table[{i=i+1;x=x+h,Y[i]},{x,-5-h,5-h,h}],
Frame->True,FrameLabel->{"x","y"},FrameTicks->All,
PlotStyle->{Black},PlotRange->{0,80}];

p2=Plot[{Cosh[x]},{x,-5,5},PlotStyle->{Black},
PlotRange->{0,80}];

Show[p1,p2]
```

Figures 1.43, 1.44, 1.45, 1.46, 1.47, 1.48, and 1.49 show numerical solution of differential equation obeyed by cosh(x) obtained using Taylor series method using various values of increment h in the interval 0.25 to 1.75, using program number 1.10. Figure 1.43 shows data of Table 1.7. For initial values $(x, v) = (-3, -10.0179)$ and $(x, y) = (-3, 10.0677)$. The curve is analytic solution given by $y = \cosh(x) = \dfrac{e^x + e^{-x}}{2}$. Agreement between numerical and analytical solutions remain perfect for increment h up to 1.25 and deteriorates for larger values of h.

Figures 1.50, 1.51, 1.52, 1.53, 1.54, 1.55, and 1.56 show numerical solution of differential equation obeyed by $\cosh(x)$ obtained using Taylor series method using various values of increment h in the interval 0.25 to 1.75, using program number 1.11. Figure 1.50 shows data of Table 1.8. For initial values $(x, v) = (-5, -74.2032)$ and $(x, y) = (-5, 74.2099)$. The curve is analytic solution given by $y = \cosh(x) = \dfrac{e^x + e^{-x}}{2}$. Agreement between numerical and analytical results remain perfect for values of increment h up to 1.00 and deteriorates for larger values of h.

We have numerically solved differential equation obeyed by Hyperbolic function cosh as an initial value problem using Taylor series method. We find that we can use large values of increment of the independent variable and get numerical solutions that closely match with known exact analytic solution. We have performed symbolic computation in Mathematica® using which we have got the numerical solutions.

Numerical Solutions Using Taylor Series Method: Boundary Value Problems

2

2.1 Reproducing Hermite Polynomial H_3 Using Shooting Method Involving Taylor Series Method

The Hermite differential equation is

$$\frac{d^2 y}{dx^2} - 2x\frac{dy}{dx} + 2ny = 0 \tag{2.1}$$

This has well known polynomial solutions for integer values of n. For $n = 3$, the solution is known to be $H_3 = 3x - 2x^3$. Here with two given boundary values: $y_a = H_3\,(x = -2) = +10$ and $y_b = H_3\,(x = +2) = -10$, we wish to solve the differential equation using Taylor series method using the boundary value y_a as one initial value. The other initial value is the value of $V = \dfrac{dy}{dx}$ for $x = -2$ which we will find out by trial and error. We are assuming that the function $H_3 = 3x - 2x^3$ is unknown. Correct initial value of V will yield correct value of y for $x = +2$. This is Shooting method involving Taylor series method.

Taylor series is known to be

$$y(x+h) = y(x) + (h/1!)\,y^{\mathrm{I}}(x) + (h^2/2!)\,y^{\mathrm{II}}(x)$$
$$+ (h^3/3!)\,y^{\mathrm{III}}(x) + (h^4/4!)\,y^{\mathrm{IV}}(x) + (h^5/5!)\,y^{\mathrm{V}}(x)$$

retaining terms containing upto fifth derivative. As to the Hermite differential equation $\dfrac{d^2 y}{dx^2} - 2x\dfrac{dy}{dx} + 2ny = 0$ with $n = 3$, we have

$$V^{\mathrm{I}} = 2xV - 6y \tag{2.2}$$

© The Author(s), under exclusive license to Springer Nature Switzerland AG 2026
S. Chowdhury, M. G. Moktadir, *Numerical Solutions Using the Taylor Series Method*, Synthesis Lectures on Mathematics & Statistics,
https://doi.org/10.1007/978-3-032-26992-8_2

and hence

$$V^{I} = 2xV - 6y$$

$$V^{II} = 2xV^{I} - 4V$$

$$V^{III} = 2xV^{II} - 2V^{I}$$

$$V^{IV} = 2xV^{III}$$

$$V^{V} = 2xV^{IV} + 2V^{III}$$

where

$$\frac{dy}{dx} = V \tag{2.3}$$

and V^{I}, V^{II}, V^{III}, V^{IV}, V^{V} are first, second, third, fourth and fifth derivatives of V with respect to x respectively; we denote these as vd1, vd2, vd3, vd4, vd5 respectively.

Taylor series solution for Eq. (2.2): $V^{I} = 2xV - 6y$ is given by

$$V_{new} = V_{old} + (h/1!)V^{I} + (h^{2}/2!)V^{II} + (h^{3}/3!)V^{III} + (h^{4}/4!)V^{IV} + (h^{5}/5!)V^{V}$$

retaining terms up to fifth derivative. And Taylor series solution for Eq. (2.3): $\dfrac{dy}{dx} = V$ is given by

$$y_{new} = y_{old} + (h/1!)y^{I} + (h^{2}/2!)y^{II} + (h^{3}/3!)y^{III} + (h^{4}/4!)y^{IV} + (h^{5}/5!)y^{V}$$

retaining terms up to fifth derivative. Here

$$y^{I} = V$$

$$y^{II} = V^{I}$$

$$y^{III} = V^{II}$$

$$y^{IV} = V^{III}$$

$$y^{V} = V^{IV}$$

which we denote by yd1, yd2, yd3, yd4, yd5 respectively.

We have written program number 2.1 to obtain values of y for $x = +2$ using the given value y_a as the value of y for $x = -2$. We have run the program for various chosen values of V at $x = -2$. Accordingly we have obtained the data tabulated as Tables 2.1, 2.2, 2.3, 2.4, 2.5, 2.6, and 2.7 (See Figs. 2.1, 2.2, 2.3, 2.4, 2.5, 2.6, and 2.7). From the bottom rows of these seven tables, we have gathered values of y for $x = +2$ and we have denoted these as y_r shown in Table 2.8 (See Figs. 2.8 and 2.9).

Table 2.1 Numerical solution of Hermite differential equation for $n = 3$ for $y_a = +10$ and chosen value of $V = 0$ at $x = -2$. Using program number 2.1

i	x	V	y
1	−1.9	−4.9273	9.7368
2	−1.8	−8.1307	9.0718
3	−1.7	−10.0913	8.1519
4	−1.6	−11.1439	7.0837
5	−1.5	−11.5265	5.9454
6	−1.4	−11.4115	4.7950
7	−1.3	−10.9266	3.6754
8	−1.2	−10.1681	2.6187
9	−1.1	−9.2101	1.6484
10	−1	−8.1110	0.7814
11	−0.9	−6.9174	0.0294
12	−0.8	−5.6676	−0.6002
13	−0.7	−4.3933	−1.1033
14	−0.6	−3.1218	−1.4790
15	−0.5	−1.8767	−1.7286
16	−0.4	−0.6788	−1.8559
17	−0.3	0.4529	−1.8665
18	−0.2	1.5011	−1.7681
19	−0.1	2.4492	−1.5697
20	0	3.2815	−1.2821
21	0.1	3.9826	−0.9177
22	0.2	4.5369	−0.4905
23	0.3	4.9287	−0.0158
24	0.4	5.1417	0.4893
25	0.5	5.1582	1.0060
26	0.6	4.9595	1.5138
27	0.7	4.5246	1.9901
28	0.8	3.8299	2.4100
29	0.9	2.8483	2.7465
30	1	1.5480	2.9691
31	1.1	−0.1094	3.0442
32	1.2	−2.1704	2.9338
33	1.3	−4.6934	2.5947
34	1.4	−7.7526	1.9773
35	1.5	−11.4441	1.0232
36	1.6	−15.8957	−0.3369
37	1.7	−21.2799	−2.1870
38	1.8	−27.8345	−4.6317
39	1.9	−35.8945	−7.8039
40	2	−45.9407	−11.8766

Table 2.2 Numerical solution of Hermite differential equation for $n = 3$ for $y_a = +10$ and chosen value of $V = -10$ at $x = -2$. Using program number 2.1

i	x	V	y
1	−1.9	−11.4667	8.9183
2	−1.8	−12.0875	7.7348
3	−1.7	−12.1145	6.5205
4	−1.6	−11.7230	5.3258
5	−1.5	−11.0377	4.1857
6	−1.4	−10.1489	3.1250
7	−1.3	−9.1235	2.1605
8	−1.2	−8.0119	1.3032
9	−1.1	−6.8529	0.5597
10	−1	−5.6772	−0.0669
11	−0.9	−4.5091	−0.5760
12	−0.8	−3.3687	−0.9696
13	−0.7	−2.2727	−1.2513
14	−0.6	−1.2352	−1.4261
15	−0.5	−0.2687	−1.5007
16	−0.4	0.6159	−1.4826
17	−0.3	1.4087	−1.3806
18	−0.2	2.1006	−1.2042
19	−0.1	2.6829	−0.9641
20	0	3.1475	−0.6716
21	0.1	3.4861	−0.3388
22	0.2	3.6908	0.0212
23	0.3	3.7532	0.3946
24	0.4	3.6647	0.7668
25	0.5	3.4162	1.1222
26	0.6	2.9978	1.4444
27	0.7	2.3986	1.7157
28	0.8	1.6061	1.9176
29	0.9	0.6063	2.0301
30	1	−0.6177	2.0314
31	1.1	−2.0859	1.8984
32	1.2	−3.8226	1.6053
33	1.3	−5.8585	1.1239
34	1.4	−8.2323	0.4224
35	1.5	−10.9945	−0.5355
36	1.6	−14.2121	−1.7917
37	1.7	−17.9752	−3.3961
38	1.8	−22.4086	−5.4090
39	1.9	−27.6876	−7.9059
40	2	−34.0642	−10.9830

Table 2.3 Numerical solution of Hermite differential equation for $n = 3$ for $y_a = +10$ and chosen value of $V = -20$ at $x = -2$. Using program number 2.1

i	x	V	y
1	−1.9	−18.0061	8.0999
2	−1.8	−16.0443	6.3977
3	−1.7	−14.1377	4.8891
4	−1.6	−12.3021	3.5678
5	−1.5	−10.5489	2.4260
6	−1.4	−8.8863	1.4550
7	−1.3	−7.3203	0.6455
8	−1.2	−5.8556	−0.0124
9	−1.1	−4.4957	−0.5291
10	−1	−3.2434	−0.9152
11	−0.9	−2.1008	−1.1815
12	−0.8	−1.0699	−1.3391
13	−0.7	−0.1521	−1.3992
14	−0.6	0.6513	−1.3733
15	−0.5	1.3392	−1.2728
16	−0.4	1.9105	−1.1093
17	−0.3	2.3644	−0.8946
18	−0.2	2.7001	−0.6404
19	−0.1	2.9166	−0.3586
20	0	3.0134	−0.0611
21	0.1	2.9897	0.2401
22	0.2	2.8446	0.5328
23	0.3	2.5776	0.8050
24	0.4	2.1877	1.0443
25	0.5	1.6742	1.2384
26	0.6	1.0362	1.3749
27	0.7	0.2726	1.4414
28	0.8	−0.6176	1.4252
29	0.9	−1.6358	1.3136
30	1	−2.7834	1.0938
31	1.1	−4.0624	0.7526
32	1.2	−5.4748	0.2768
33	1.3	−7.0235	−0.3469
34	1.4	−8.7120	−1.1325
35	1.5	−10.5450	−2.0941
36	1.6	−12.5284	−3.2465
37	1.7	−14.6705	−4.6051
38	1.8	−16.9826	−6.1863
39	1.9	−19.4807	−8.0078
40	2	−22.1877	−10.0894

Table 2.4 Numerical solution of Hermite differential equation for $n = 3$ for $y_a = +10$ and chosen value of $V = -30$ at $x = -2$. Using program number 2.1

i	x	V	y
1	−1.9	−24.5455	7.2814
2	−1.8	−20.0011	5.0607
3	−1.7	−16.1609	3.2578
4	−1.6	−12.8812	1.8098
5	−1.5	−10.0601	0.6663
6	−1.4	−7.6237	−0.2150
7	−1.3	−5.5172	−0.8695
8	−1.2	−3.6994	−1.3280
9	−1.1	−2.1385	−1.6179
10	−1	−0.8096	−1.7635
11	−0.9	0.3075	−1.7869
12	−0.8	1.2290	−1.7085
13	−0.7	1.9686	−1.5471
14	−0.6	2.5379	−1.3205
15	−0.5	2.9472	−1.0449
16	−0.4	3.2052	−0.7361
17	−0.3	3.3202	−0.4086
18	−0.2	3.2995	−0.0765
19	−0.1	3.1503	0.2470
20	0	2.8794	0.5495
21	0.1	2.4932	0.8190
22	0.2	1.9985	1.0445
23	0.3	1.4020	1.2153
24	0.4	0.7107	1.3217
25	0.5	−0.0678	1.3546
26	0.6	−0.9255	1.3055
27	0.7	−1.8534	1.1671
28	0.8	−2.8414	0.9328
29	0.9	−3.8779	0.5972
30	1	−4.9491	0.1561
31	1.1	−6.0388	−0.3932
32	1.2	−7.1270	−1.0516
33	1.3	−8.1885	−1.8177
34	1.4	−9.1918	−2.6874
35	1.5	−10.0954	−3.6528
36	1.6	−10.8447	−4.7014
37	1.7	−11.3658	−5.8141
38	1.8	−11.5566	−6.9635
39	1.9	−11.2738	−8.1098
40	2	−10.3111	−9.1958

Table 2.5 Numerical solution of Hermite differential equation for $n = 3$ for $y_a = +10$ and chosen value of $V = -40$ at $x = -2$. Using program number 2.1

i	x	V	y
1	−1.9	−31.0849	6.4629
2	−1.8	−23.9579	3.7236
3	−1.7	−18.1840	1.6264
4	−1.6	−13.4603	0.0519
5	−1.5	−9.5713	−1.0935
6	−1.4	−6.3610	−1.8850
7	−1.3	−3.7140	−2.3844
8	−1.2	−1.5432	−2.6436
9	−1.1	0.2187	−2.7067
10	−1	1.6243	−2.6118
11	−0.9	2.7158	−2.3923
12	−0.8	3.5278	−2.0779
13	−0.7	4.0892	−1.6951
14	−0.6	4.4245	−1.2676
15	−0.5	4.5551	−0.8170
16	−0.4	4.4999	−0.3628
17	−0.3	4.2759	0.0773
18	−0.2	3.8990	0.4873
19	−0.1	3.3841	0.8525
20	0	2.7453	1.1600
21	0.1	1.9967	1.3979
22	0.2	1.1523	1.5561
23	0.3	0.2264	1.6257
24	0.4	−0.7663	1.5992
25	0.5	−1.8098	1.4707
26	0.6	−2.8871	1.2361
27	0.7	−3.9794	0.8928
28	0.8	−5.0652	0.4404
29	0.9	−6.1199	−0.1192
30	1	−7.1148	−0.7816
31	1.1	−8.0153	−1.5390
32	1.2	−8.7792	−2.3801
33	1.3	−9.3536	−3.2886
34	1.4	−9.6715	−4.2423
35	1.5	−9.6458	−5.2114
36	1.6	−9.1610	−6.1562
37	1.7	−8.0611	−7.0232
38	1.8	−6.1307	−7.7408
39	1.9	−3.0669	−8.2117
40	2	1.5654	−8.3021

Table 2.6 Numerical solution of Hermite differential equation for $n = 3$ for $y_a = +10$ and chosen value of $V = -50$ at $x = -2$. Using program number 2.1

i	x	V	y
1	−1.9	−37.6243	5.6445
2	−1.8	−27.9147	2.3866
3	−1.7	−20.2072	−0.0050
4	−1.6	−14.0393	−1.7061
5	−1.5	−9.0825	−2.8532
6	−1.4	−5.0984	−3.5550
7	−1.3	−1.9109	−3.8994
8	−1.2	0.6131	−3.9592
9	−1.1	2.5759	−3.7955
10	−1	4.0581	−3.4600
11	−0.9	5.1241	−2.9977
12	−0.8	5.8266	−2.4473
13	−0.7	6.2098	−1.8430
14	−0.6	6.3111	−1.2148
15	−0.5	6.1630	−0.5891
16	−0.4	5.7945	0.0105
17	−0.3	5.2317	0.5633
18	−0.2	4.4985	1.0511
19	−0.1	3.6178	1.4581
20	0	2.6112	1.7705
21	0.1	1.5003	1.9769
22	0.2	0.3062	2.0678
23	0.3	−0.9492	2.0361
24	0.4	−2.2432	1.8767
25	0.5	−3.5518	1.5869
26	0.6	−4.8488	1.1667
27	0.7	−6.1054	0.6185
28	0.8	−7.2889	−0.0520
29	0.9	−8.3620	−0.8356
30	1	−9.2805	−1.7192
31	1.1	−9.9918	−2.6848
32	1.2	−10.4313	−3.7086
33	1.3	−10.5186	−4.7594
34	1.4	−10.1512	−5.7972
35	1.5	−9.1962	−6.7701
36	1.6	−7.4773	−7.6110
37	1.7	−4.7564	−8.2322
38	1.8	−0.7047	−8.5181
39	1.9	5.1401	−8.3137
40	2	13.4419	−7.4085

Table 2.7 Numerical solution of Hermite differential equation for $n = 3$ for $y_a = +10$ and chosen value of $V = -60$ at $x = -2$. Using program number 2.1

i	x	V	y
1	−1.9	−44.1637	4.8260
2	−1.8	−31.8715	1.0495
3	−1.7	−22.2304	−1.6364
4	−1.6	−14.6184	−3.4640
5	−1.5	−8.5937	−4.6129
6	−1.4	−3.8358	−5.2249
7	−1.3	−0.1078	−5.4144
8	−1.2	2.7693	−5.2748
9	−1.1	4.9331	−4.8842
10	−1	6.4919	−4.3083
11	−0.9	7.5323	−3.6031
12	−0.8	8.1255	−2.8168
13	−0.7	8.3304	−1.9910
14	−0.6	8.1977	−1.1619
15	−0.5	7.7710	−0.3612
16	−0.4	7.0892	0.3838
17	−0.3	6.1874	1.0493
18	−0.2	5.0980	1.6150
19	−0.1	3.8515	2.0637
20	0	2.4772	2.3810
21	0.1	1.0038	2.5558
22	0.2	−0.5400	2.5794
23	0.3	−2.1248	2.4464
24	0.4	−3.7202	2.1541
25	0.5	−5.2938	1.7031
26	0.6	−6.8104	1.0973
27	0.7	−8.2313	0.3442
28	0.8	−9.5127	−0.5444
29	0.9	−10.6040	−1.5520
30	1	−11.4462	−2.6569
31	1.1	−11.9683	−3.8306
32	1.2	−12.0835	−5.0370
33	1.3	−11.6836	−6.2302
34	1.4	−10.6310	−7.3520
35	1.5	−8.7466	−8.3287
36	1.6	−5.7936	−9.0658
37	1.7	−1.4516	−9.4413
38	1.8	4.7213	−9.2953
39	1.9	13.3470	−8.4156
40	2	25.3185	−6.5149

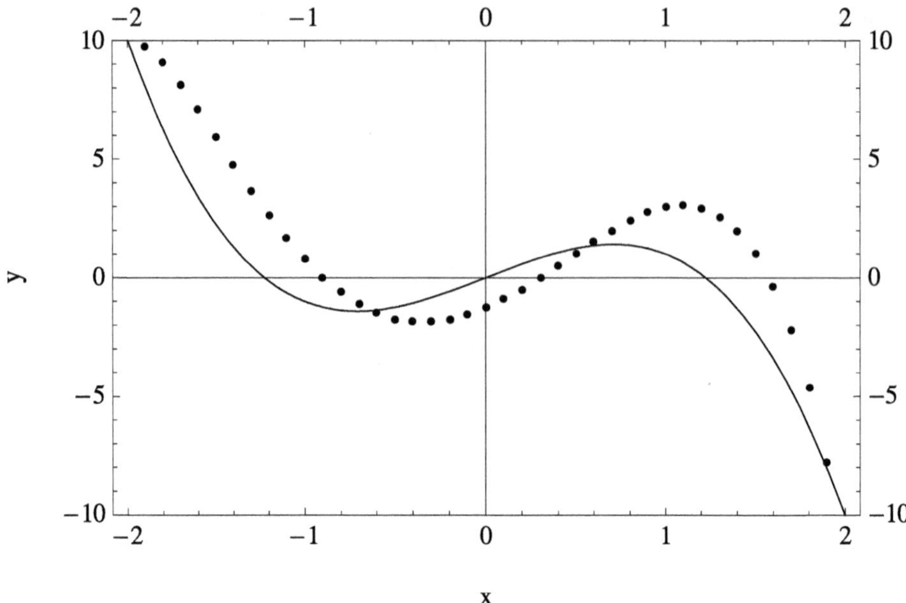

Fig. 2.1 Showing data of Table 2.1 as numerical solution of Hermite differential equation for $n = 3$ for $y_a = +10$ and chosen value of $V = 0$ for $x = -2$. The curve is for known exact result. Using program number 2.1

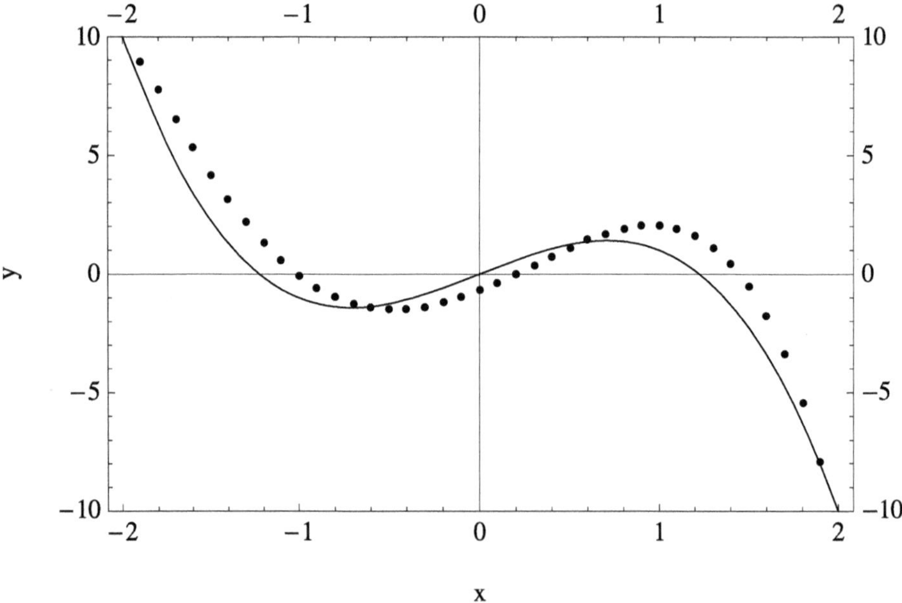

Fig. 2.2 Showing data of Table 2.2 as numerical solution of Hermite differential equation for $n = 3$ for $y_a = +10$ and chosen value of $V = -10$ for $x = -2$. The curve is for known exact result. Using program number 2.1

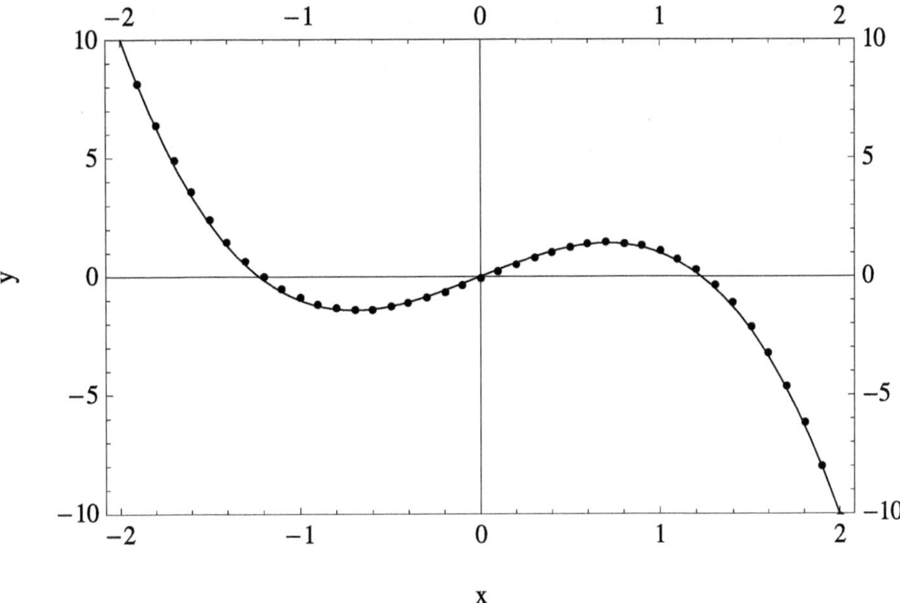

Fig. 2.3 Showing data of Table 2.3 as numerical solution of Hermite differential equation for $n = 3$ for $y_a = +10$ and chosen value of $V = -20$ for $x = -2$. The curve is for known exact result. Using program number 2.1

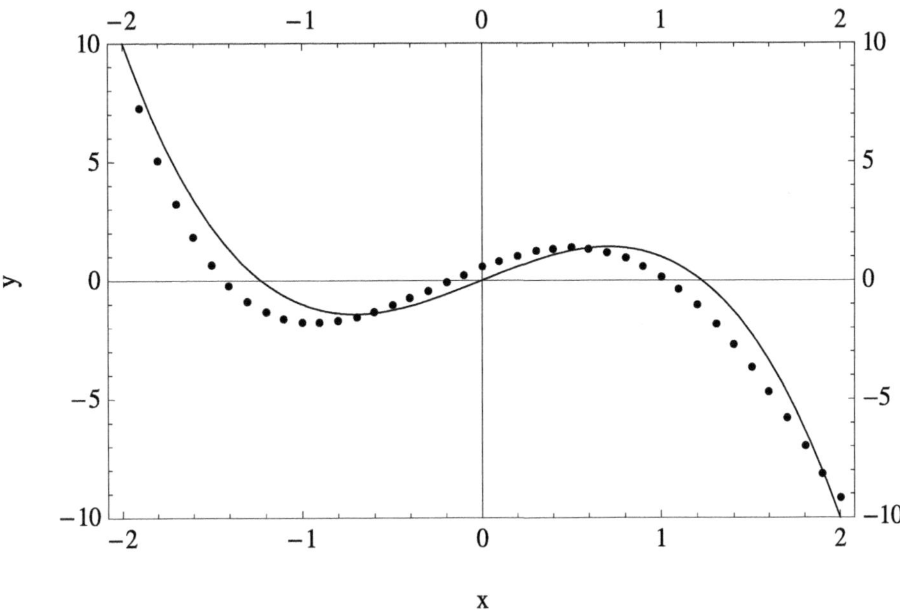

Fig. 2.4 Showing data of Table 2.4 as numerical solution of Hermite differential equation for $n = 3$ for $y_a = +10$ and chosen value of $V = -30$ for $x = -2$. The curve is for known exact result. Using program number 2.1

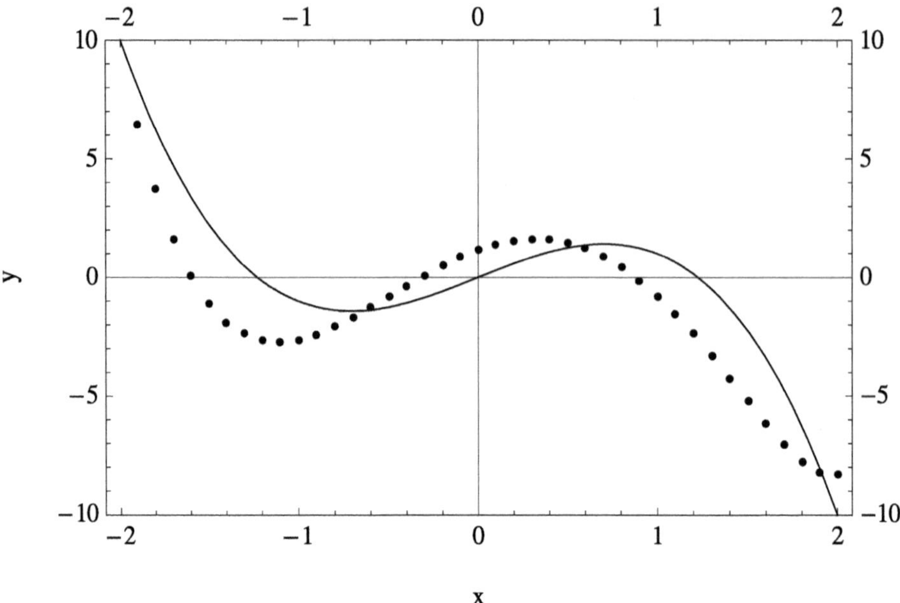

Fig. 2.5 Showing data of Table 2.5 as numerical solution of Hermite differential equation for $n = 3$ for $y_a = +10$ and chosen value of $V = -40$ for $x = -2$. The curve is for known exact result. Using program number 2.1

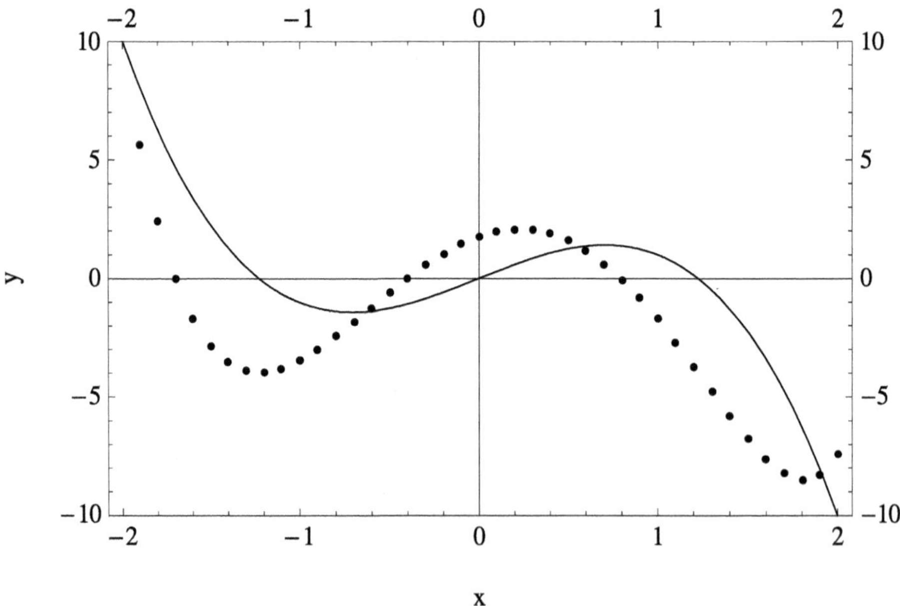

Fig. 2.6 Showing data of Table 2.6 as numerical solution of Hermite differential equation for $n = 3$ for $y_a = +10$ and chosen value of $V = -50$ for $x = -2$. The curve is for known exact result. Using program number 2.1

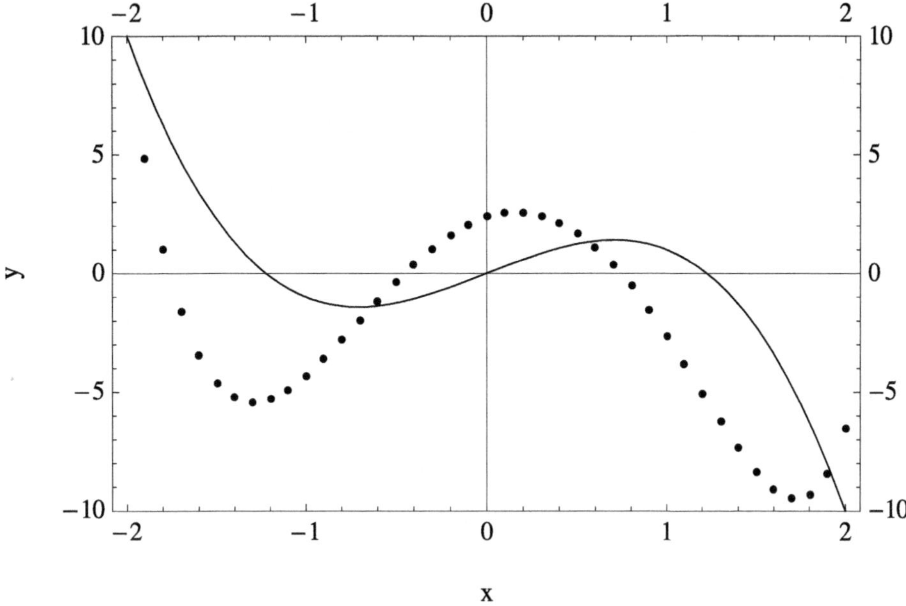

Fig. 2.7 Showing data of Table 2.7 as numerical solution of Hermite differential equation for $n = 3$ for $y_a = +10$ and chosen value of $V = -60$ for $x = -2$. The curve is for known exact result. Using program number 2.1

Table 2.8 Showing values of y called y_r gathered from last rows of Tables 2.1, 2.2, 2.3, 2.4, 2.5, 2.6, and 2.7. We have values of y_r for various chosen values of V at $x = -2$ in the interval 0 to −60

V	y_r	$d = y_b - y_r = -10 - y_r$
0	−11.8766	1.8766
−10	−10.983	0.983
−20	−10.0894	0.0894
−30	−9.1958	−0.8042
−40	−8.3021	−1.6979
−50	−7.4085	−2.5915
−60	−6.5149	−3.4851

Program Number 2.1 (for Hermite polynomial H_3)

```
  h=0.1;
x=-2;
y=10;
v=-21;
i=0;
Table [{

i=i+1,
x=x+h,
```

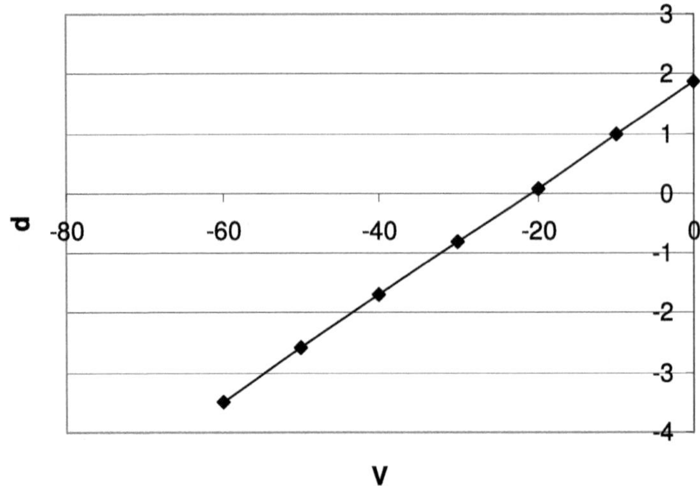

Fig. 2.8 Showing $d = y_b - y_r = -10 - y_r$ of Table 2.8 as a function of various chosen values of V at $x = -2$. We find that for $V = -21$, we have $d = 0$ i.e. $y_b = y_r$

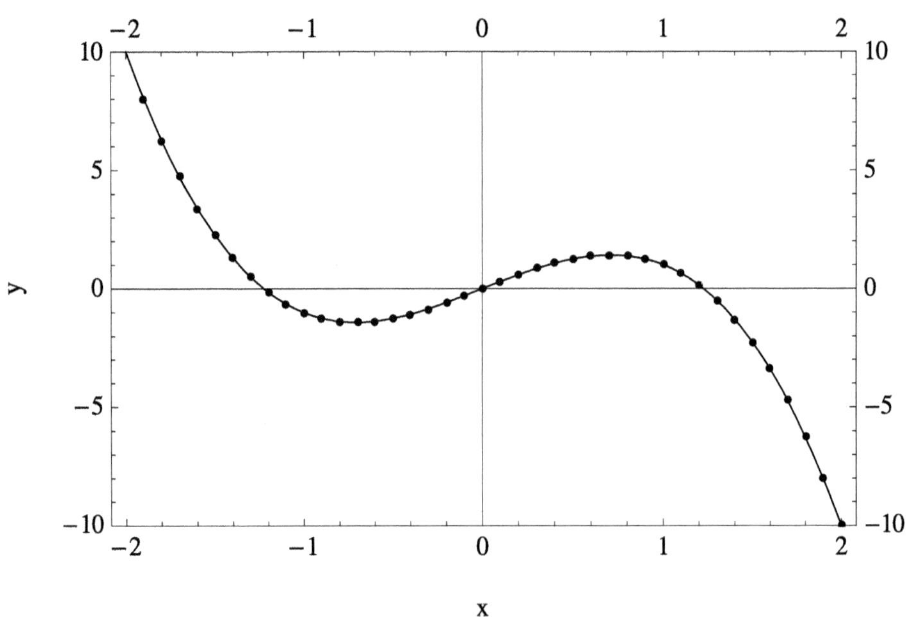

Fig. 2.9 Showing numerical solution of Hermite differential equation for $n = 3$ for given $y_a = +10$ and $V = -21$ at $x = -2$ obtained using Shooting method using program number 2.1. The curve is for known exact result: $y = H_3 = 3x - 2x^3$; agreement is perfect

```
vd1=2*(x-h)*v-6*y;
vd2=2*(x-h)*vd1-4*v;
vd3=2*(x-h)*vd2-2*vd1;
```

```
vd4=2*(x-h)*vd3;
vd5=2*(x-h)*vd4+2*vd3;

yd1=v;
yd2=vd1;
yd3=vd2;
yd4=vd3;
yd5=vd4;

v=v+h*vd1+(1/2)*h^2*vd2+(1/6)*h^3*vd3+
(1/24)*h^4*vd4+(1/120)*h^5*vd5,

Y[i]=y=y+h*yd1+(1/2)*h^2*yd2+(1/6)*h^3*yd3+
(1/24)*h^4*yd4+(1/120)*h^5*yd5},

{x,-2,2-h,h}];
TableForm[%,TableSpacing->{2,2},
TableHeadings->{None,{"i","x","V","y"}}]

i=0;
p1=ListPlot[Table[{i=i+1;x=x+h,Y[i]},{x,-2,2-h,h}],
Frame->True,FrameLabel->{"x","y"},FrameTicks->All,
PlotStyle->{Black},PlotRange->{-10,10}];

p2=Plot[{3*x-2*x^3},{x,-2,2},
PlotStyle->{Black},
PlotRange->{-10,10}];

Show[p1,p2]
```

2.2   Reproducing Hermite Polynomial H_4 Using Shooting Method Involving Taylor Series Method

The Hermite differential equation is

$$\frac{d^2 y}{dx^2} - 2x\frac{dy}{dx} + 2ny = 0$$

This has well known polynomial solutions for integer values of n. For $n = 4$, the solution is known to be $H_4 = 12{-}48x^2 + 16x^4$. We are given the two boundary values: $y_a = H_4$ $(x = -2) = +76$ and $y_b = H_4$ $(x = +2) = +76$. We wish to solve the differential equation using Taylor series method using the boundary value y_a as one initial value. The other initial value is the value of $V = \dfrac{dy}{dx}$ at $x = -2$ which we will find out by trial and error. We are assuming that the function $H_4 = 12 - 48x^2 + 16x^4$ is unknown. Correct initial value of V will yield correct value of y for $x = +2$. This is Shooting method involving Taylor series method.

Taylor series is known to be

$$y(x+h) = y(x) + (h/1!)y^{\mathrm{I}}(x) + (h^2/2!)y^{\mathrm{II}}(x)$$
$$+ (h^3/3!)y^{\mathrm{III}}(x) + (h^4/4!)y^{\mathrm{IV}}(x) + (h^5/5!)y^{\mathrm{V}}(x)$$

retaining terms containing upto fifth derivative. As to the Hermite differential equation $\dfrac{d^2 y}{dx^2} - 2x\dfrac{dy}{dx} + 2ny = 0$ with $n = 4$, we have

$$V^{\mathrm{I}} = 2xV - 8y \tag{2.4}$$

and hence

$$V^{\mathrm{I}} = 2xV - 8y$$

$$V^{\mathrm{II}} = 2xV^{\mathrm{I}} - 6V$$

$$V^{\mathrm{III}} = 2xV^{\mathrm{II}} - 4V^{\mathrm{I}}$$

$$V^{\mathrm{IV}} = 2xV^{\mathrm{III}} - 2V^{\mathrm{II}}$$

$$V^{\mathrm{V}} = 2xV^{\mathrm{IV}}$$

where

$$\frac{dy}{dx} = V \tag{2.5}$$

and V^{I}, V^{II}, V^{III}, V^{IV}, V^{V} are first, second, third, fourth and fifth derivatives of V with respect to x respectively; we denote these as vd1, vd2, vd3, vd4, vd5 respectively.

Taylor series solution for Eq. (2.4): $V^{\mathrm{I}} = 2xV - 8y$ is given by

$$V_{\mathrm{new}} = V_{\mathrm{old}} + (h/1!)V^{\mathrm{I}} + (h^2/2!)V^{\mathrm{II}} + (h^3/3!)V^{\mathrm{III}} + (h^4/4!)V^{\mathrm{IV}} + (h^5/5!)V^{\mathrm{V}}$$

retaining terms up to fifth derivative. And Taylor series solution for Eq. (2.5): $\dfrac{dy}{dx} = V$ is given by

$$y_{\mathrm{new}} = y_{\mathrm{old}} + (h/1!)y^{\mathrm{I}} + (h^2/2!)y^{\mathrm{II}} + (h^3/3!)y^{\mathrm{III}} + (h^4/4!)y^{\mathrm{IV}} + (h^5/5!)y^{\mathrm{V}}$$

retaining terms up to fifth derivative. Here

$$y^{\mathrm{I}} = V$$

$$y^{\mathrm{II}} = V^{\mathrm{I}}$$

$$y^{\mathrm{III}} = V^{\mathrm{II}}$$

$$y^{\mathrm{IV}} = V^{\mathrm{III}}$$

$$y^{\mathrm{V}} = V^{\mathrm{IV}}$$

which we denote by yd1, yd2, yd3, yd4, yd5 respectively.

We have written program number 2.2 to obtain values of y for $x = +2$ using the given value y_a as the value of y for $x = -2$. We have run the program for various chosen values of V at $x = -2$. Accordingly we have obtained the data tabulated as Tables 2.9, 2.10, 2.11, 2.12, 2.13, 2.14, and 2.15 (See Figs. 2.10, 2.11, 2.12, 2.13, 2.14, 2.15, and 2.16). From the bottom rows of these seven tables, we have gathered values of y for $x = +2$ and we have denoted these as y_r shown in Table 2.16 (See Figs. 2.17 and 2.18).

Program Number 2.2 (for Hermite polynomial H_4)

```
  h=0.1;
x=-2;
y=76;
v=-320;
i=0;

Table[{
i=i+1,
x=x+h,

vd1=2*(x-h)*v-8*y;
vd2=2*(x-h)*vd1-6*v;
vd3=2*(x-h)*vd2-4*vd1;
vd4=2*(x-h)*vd3-2*vd2;
vd5=2*(x-h)*vd4;

yd1=v;
yd2=vd1;
yd3=vd2;
yd4=vd3;
yd5=vd4;

v=v+h*vd1+(1/2)*h^2*vd2+(1/6)*h^3*vd3+
(1/24)*h^4*vd4+(1/120)*h^5*vd5,

Y[i]=y=y+h*yd1+(1/2)*h^2*yd2+(1/6)*h^3*yd3+
(1/24)*h^4*yd4+(1/120)*h^5*yd5},
```

Table 2.9 Numerical solution of Hermite differential equation for $n = 4$ for $y_a = +76$ and chosen value of $V = 0$ at $x = -2$. Using program number 2.2

i	x	V	y
1	−1.9	−49.7628	73.3370
2	−1.8	−81.2928	66.6536
3	−1.7	−99.1891	57.5318
4	−1.6	−106.8840	47.1550
5	−1.5	−106.9920	36.4072
6	−1.4	−101.5360	25.9417
7	−1.3	−92.1104	16.2321
8	−1.2	−79.9888	7.6093
9	−1.1	−66.1985	0.2898
10	−1	−51.5768	−5.6029
11	−0.9	−36.8099	−10.0209
12	−0.8	−22.4617	−12.9789
13	−0.7	−8.9951	−14.5427
14	−0.6	3.2122	−14.8200
15	−0.5	13.8556	−13.9525
16	−0.4	22.6949	−12.1091
17	−0.3	29.5470	−9.4798
18	−0.2	34.2818	−6.2703
19	−0.1	36.8177	−2.6968
20	0	37.1199	1.0187
21	0.1	35.1986	4.6530
22	0.2	31.1087	7.9861
23	0.3	24.9505	10.8058
24	0.4	16.8704	12.9121
25	0.5	7.0640	14.1223
26	0.6	−4.2206	14.2756
27	0.7	−16.6778	13.2390
28	0.8	−29.9380	10.9132
29	0.9	−43.5581	7.2393
30	1	−57.0101	2.2070
31	1.1	−69.6649	−4.1363
32	1.2	−80.7712	−11.6745
33	1.3	−89.4264	−20.2091
34	1.4	−94.5376	−29.4419
35	1.5	−94.7669	−38.9542
36	1.6	−88.4548	−48.1778
37	1.7	−73.5116	−56.3582
38	1.8	−47.2591	−62.5041
39	1.9	−6.1996	−65.3178
40	2	54.3310	−63.0969

Table 2.10 Numerical solution of Hermite differential equation for $n = 4$ for $y_a = +76$ and chosen value of $V = -100$ at $x = -2$. Using program number 2.2

i	x	V	y
1	−1.9	−114.3920	65.1797
2	−1.8	−118.5290	53.4624
3	−1.7	−115.4520	41.7136
4	−1.6	−107.4030	30.5371
5	−1.5	−96.0568	20.3425
6	−1.4	−82.6858	11.3929
7	−1.3	−68.2659	3.8401
8	−1.2	−53.5523	−2.2506
9	−1.1	−39.1315	−6.8803
10	−1	−25.4590	−10.1020
11	−0.9	−12.8868	−12.0089
12	−0.8	−1.6824	−12.7250
13	−0.7	7.9559	−12.3976
14	−0.6	15.8884	−11.1907
15	−0.5	22.0258	−9.2798
16	−0.4	26.3227	−6.8470
17	−0.3	28.7736	−4.0769
18	−0.2	29.4087	−1.1528
19	−0.1	28.2922	1.7465
20	0	25.5199	4.4504
21	0.1	21.2190	6.7995
22	0.2	15.5473	8.6485
23	0.3	8.6934	9.8695
24	0.4	0.8784	10.3551
25	0.5	−7.6435	10.0215
26	0.6	−16.5816	8.8125
27	0.7	−25.6060	6.7023
28	0.8	−34.3423	3.7008
29	0.9	−42.3662	−0.1425
30	1	−49.1944	−4.7327
31	1.1	−54.2746	−9.9232
32	1.2	−56.9702	−15.5082
33	1.3	−56.5407	−21.2132
34	1.4	−52.1146	−26.6833
35	1.5	−42.6522	−31.4685
36	1.6	−26.8927	−35.0042
37	1.7	−3.2792	−36.5857
38	1.8	30.1493	−35.3335
39	1.9	75.9178	−30.1455
40	2	137.3530	−19.6291

Table 2.11 Numerical solution of Hermite differential equation for $n = 4$ for $y_a = +76$ and chosen value of $V = -200$ at $x = -2$. Using program number 2.2

i	x	V	y
1	−1.9	−179.0210	57.0224
2	−1.8	−155.7650	40.2711
3	−1.7	−131.7160	25.8954
4	−1.6	−107.9220	13.9191
5	−1.5	−85.1219	4.2777
6	−1.4	−63.8359	−3.1559
7	−1.3	−44.4214	−8.5520
8	−1.2	−27.1158	−12.1105
9	−1.1	−12.0644	−14.0503
10	−1	0.6587	−14.6011
11	−0.9	11.0363	−13.9968
12	−0.8	19.0969	−12.4711
13	−0.7	24.9068	−10.2525
14	−0.6	28.5646	−7.5615
15	−0.5	30.1959	−4.6072
16	−0.4	29.9506	−1.5849
17	−0.3	28.0001	1.3261
18	−0.2	24.5357	3.9646
19	−0.1	19.7666	6.1897
20	0	13.9200	7.8820
21	0.1	7.2395	8.9459
22	0.2	−0.0142	9.3108
23	0.3	−7.5636	8.9332
24	0.4	−15.1136	7.7980
25	0.5	−22.3510	5.9209
26	0.6	−28.9427	3.3493
27	0.7	−34.5342	0.1656
28	0.8	−38.7467	−3.5116
29	0.9	−41.1743	−7.5243
30	1	−41.3788	−11.6724
31	1.1	−38.8844	−15.7101
32	1.2	−33.1692	−19.3419
33	1.3	−23.6549	−22.2174
34	1.4	−9.6916	−23.9247
35	1.5	9.4624	−23.9828
36	1.6	34.6694	−21.8307
37	1.7	66.9532	−16.8133
38	1.8	107.5580	−8.1630
39	1.9	158.0350	5.0268
40	2	220.3740	23.8387

Table 2.12 Numerical solution of Hermite differential equation for $n = 4$ for $y_a = +76$ and chosen value of $V = -300$ at $x = -2$. Using program number 2.2

i	x	V	y
1	−1.9	−243.6500	48.8651
2	−1.8	−193.0010	27.0799
3	−1.7	−147.9790	10.0772
4	−1.6	−108.4400	−2.6988
5	−1.5	−74.1870	−11.7870
6	−1.4	−44.9860	−17.7046
7	−1.3	−20.5769	−20.9440
8	−1.2	−0.6793	−21.9704
9	−1.1	15.0026	−21.2204
10	−1	26.7765	−19.1002
11	−0.9	34.9594	−15.9848
12	−0.8	39.8761	−12.2172
13	−0.7	41.8578	−8.1074
14	−0.6	41.2408	−3.9323
15	−0.5	38.3660	0.0655
16	−0.4	33.5784	3.6772
17	−0.3	27.2267	6.7290
18	−0.2	19.6626	9.0821
19	−0.1	11.2411	10.6330
20	0	2.3200	11.3137
21	0.1	−6.7401	11.0923
22	0.2	−15.5757	9.9731
23	0.3	−23.8206	7.9969
24	0.4	−31.1056	5.2410
25	0.5	−37.0585	1.8201
26	0.6	−41.3038	−2.1138
27	0.7	−43.4624	−6.3711
28	0.8	−43.1511	−10.7239
29	0.9	−39.9824	−14.9060
30	1	−33.5631	−18.6121
31	1.1	−23.4941	−21.4970
32	1.2	−9.3682	−23.1757
33	1.3	9.2309	−23.2216
34	1.4	32.7314	−21.1661
35	1.5	61.5771	−16.4971
36	1.6	96.2316	−8.6571
37	1.7	137.1860	2.9591
38	1.8	184.9660	19.0075
39	1.9	240.1530	40.1991
40	2	303.3960	67.3064

Table 2.13 Numerical solution of Hermite differential equation for $n = 4$ for $y_a = +76$ and chosen value of $V = -400$ at $x = -2$. Using program number 2.2

i	x	V	y
1	−1.9	−308.2790	40.7078
2	−1.8	−230.2370	13.8886
3	−1.7	−164.2430	−5.7409
4	−1.6	−108.9590	−19.3168
5	−1.5	−63.2521	−27.8518
6	−1.4	−26.1361	−32.2534
7	−1.3	3.2676	−33.3360
8	−1.2	25.7572	−31.8303
9	−1.1	42.0696	−28.3904
10	−1	52.8942	−23.5993
11	−0.9	58.8825	−17.9728
12	−0.8	60.6554	−11.9633
13	−0.7	58.8088	−5.9623
14	−0.6	53.9169	−0.3030
15	−0.5	46.5361	4.7381
16	−0.4	37.2063	8.9393
17	−0.3	26.4532	12.1319
18	−0.2	14.7896	14.1996
19	−0.1	2.7156	15.0762
20	0	−9.2800	14.7453
21	0.1	−20.7196	13.2387
22	0.2	−31.1372	10.6355
23	0.3	−40.0776	7.0606
24	0.4	−47.0976	2.6840
25	0.5	−51.7660	−2.2806
26	0.6	−53.6649	−7.5769
27	0.7	−52.3906	−12.9077
28	0.8	−47.5555	−17.9363
29	0.9	−38.7905	−22.2878
30	1	−25.7475	−25.5518
31	1.1	−8.1038	−27.2839
32	1.2	14.4328	−27.0094
33	1.3	42.1166	−24.2257
34	1.4	75.1544	−18.4075
35	1.5	113.6920	−9.0115
36	1.6	157.7940	4.5164
37	1.7	207.4180	22.7315
38	1.8	262.3750	46.1780
39	1.9	322.2700	75.3714
40	2	386.4170	110.7740

Table 2.14 Numerical solution of Hermite differential equation for $n = 4$ for $y_a = +76$ and chosen value of $V = -500$ at $x = -2$. Using program number 2.2

i	x	V	y
1	−1.9	−372.9080	32.5505
2	−1.8	−267.4730	0.6973
3	−1.7	−180.5060	−21.5591
4	−1.6	−109.4770	−35.9347
5	−1.5	−52.3171	−43.9166
6	−1.4	−7.2862	−46.8022
7	−1.3	27.1121	−45.7280
8	−1.2	52.1937	−41.6902
9	−1.1	69.1366	−35.5605
10	−1	79.0119	−28.0984
11	−0.9	82.8056	−19.9607
12	−0.8	81.4347	−11.7094
13	−0.7	75.7597	−3.8172
14	−0.6	66.5931	3.3262
15	−0.5	54.7062	9.4108
16	−0.4	40.8341	14.2013
17	−0.3	25.6798	17.5349
18	−0.2	9.9165	19.3170
19	−0.1	−5.8100	19.5194
20	0	−20.8799	18.1770
21	0.1	−34.6992	15.3852
22	0.2	−46.6987	11.2978
23	0.3	−56.3346	6.1243
24	0.4	−63.0896	0.1269
25	0.5	−66.4735	−6.3813
26	0.6	−66.0259	−13.0400
27	0.7	−61.3187	−19.4444
28	0.8	−51.9599	−25.1487
29	0.9	−37.5986	−29.6696
30	1	−17.9318	−32.4915
31	1.1	7.2865	−33.0708
32	1.2	38.2338	−30.8431
33	1.3	75.0024	−25.2299
34	1.4	117.5770	−15.6489
35	1.5	165.8060	−1.5258
36	1.6	219.3560	17.6900
37	1.7	277.6500	42.5040
38	1.8	339.7830	73.3485
39	1.9	404.3870	110.5440
40	2	469.4390	154.2420

Table 2.15 Numerical solution of Hermite differential equation for $n = 4$ for $y_a = +76$ and chosen value of $V = -600$ at $x = -2$. Using program number 2.2

i	x	V	y
1	−1.9	−437.5380	24.3932
2	−1.8	−304.7090	−12.4939
3	−1.7	−196.7700	−37.3773
4	−1.6	−109.9960	−52.5527
5	−1.5	−41.3822	−59.9813
6	−1.4	11.5637	−61.3510
7	−1.3	50.9566	−58.1201
8	−1.2	78.6302	−51.5501
9	−1.1	96.2037	−42.7305
10	−1	105.1300	−32.5975
11	−0.9	106.7290	−21.9487
12	−0.8	102.2140	−11.4554
13	−0.7	92.7107	−1.6721
14	−0.6	79.2693	6.9555
15	−0.5	62.8763	14.0834
16	−0.4	44.4620	19.4634
17	−0.3	24.9063	22.9378
18	−0.2	5.0434	24.4345
19	−0.1	−14.3355	23.9627
20	0	−32.4799	21.6086
21	0.1	−48.6787	17.5316
22	0.2	−62.2601	11.9601
23	0.3	−72.5917	5.1879
24	0.4	−79.0816	−2.4301
25	0.5	−81.1810	−10.4820
26	0.6	−78.3870	−18.5031
27	0.7	−70.2469	−25.9811
28	0.8	−56.3643	−32.3610
29	0.9	−36.4067	−37.0514
30	1	−10.1162	−39.4312
31	1.1	22.6768	−38.8577
32	1.2	62.0348	−34.6768
33	1.3	107.8880	−26.2341
34	1.4	160.0000	−12.8903
35	1.5	217.9210	5.9599
36	1.6	280.9180	30.8636
37	1.7	347.8830	62.2764
38	1.8	417.1920	100.5190
39	1.9	486.5050	145.7160
40	2	552.4600	197.7100

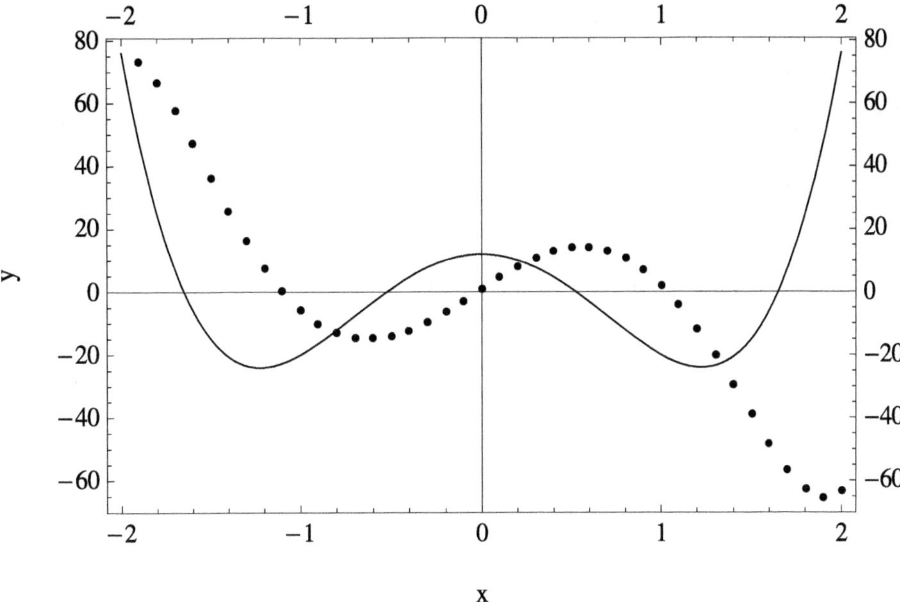

Fig. 2.10 Showing data of Table 2.9 as numerical solution of Hermite differential equation for $n = 4$ for $y_a = +76$ and chosen value of $V = 0$ for $x = -2$. The curve is for known exact result. Using program number 2.2

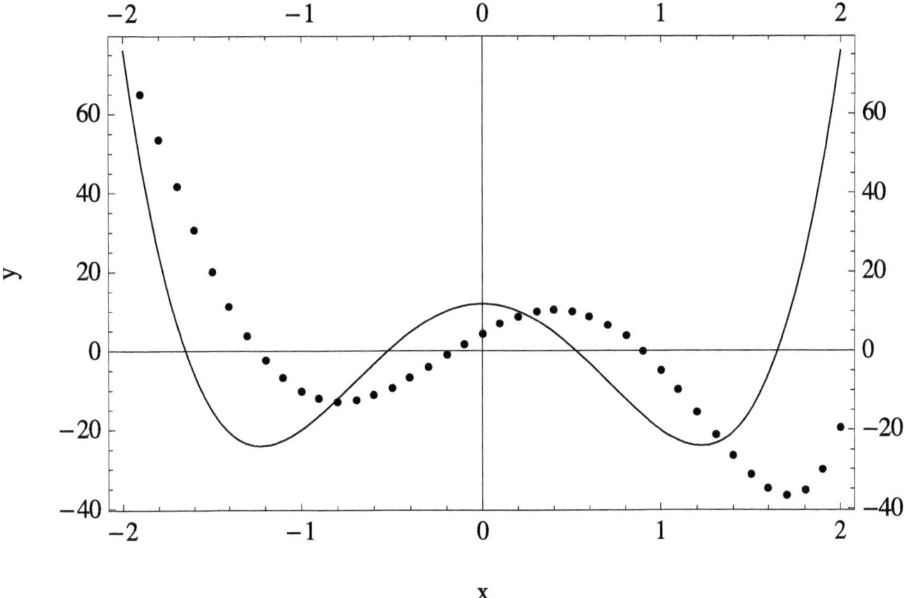

Fig. 2.11 Showing data of Table 2.10 as numerical solution of Hermite differential equation for $n = 4$ for $y_a = +76$ and chosen value of $V = -100$ for $x = -2$. The curve is for known exact result. Using program number 2.2

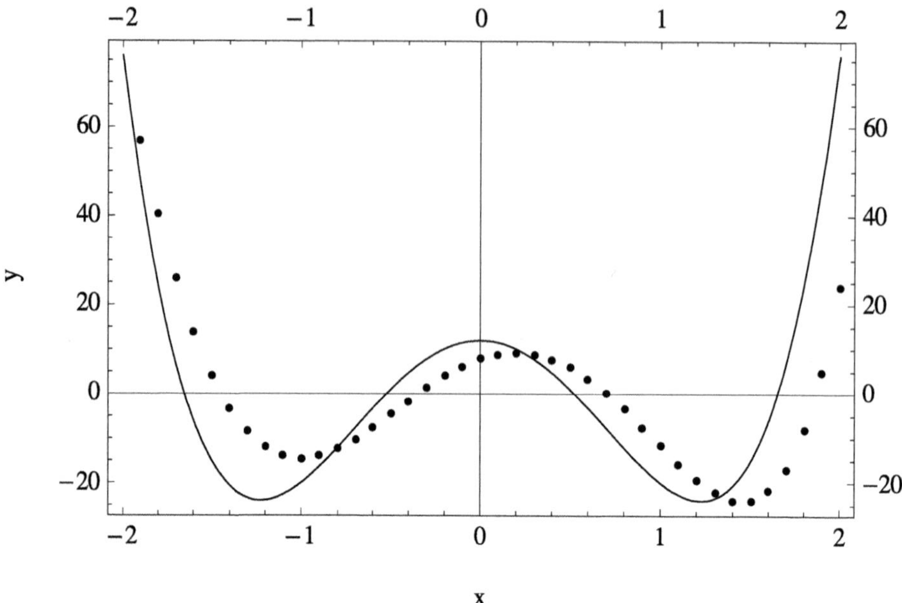

Fig. 2.12 Showing data of Table 2.11 as numerical solution of Hermite differential equation for $n = 4$ for $y_a = +76$ and chosen value of $V = -200$ for $x = -2$. The curve is for known exact result. Using program number 2.2

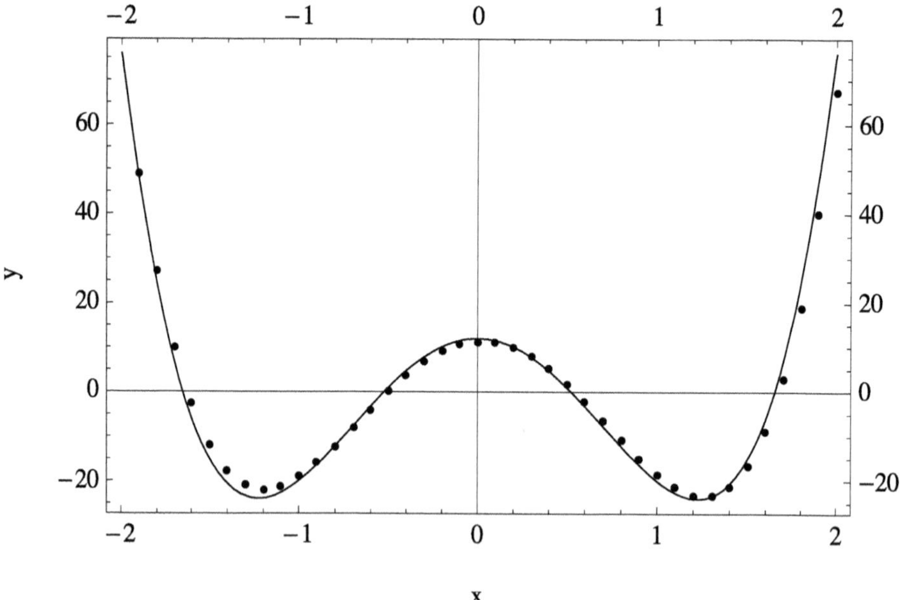

Fig. 2.13 Showing data of Table 2.12 as numerical solution of Hermite differential equation for $n = 4$ for $y_a = +76$ and chosen value of $V = -300$ for $x = -2$. The curve is for known exact result. Using program number 2.2

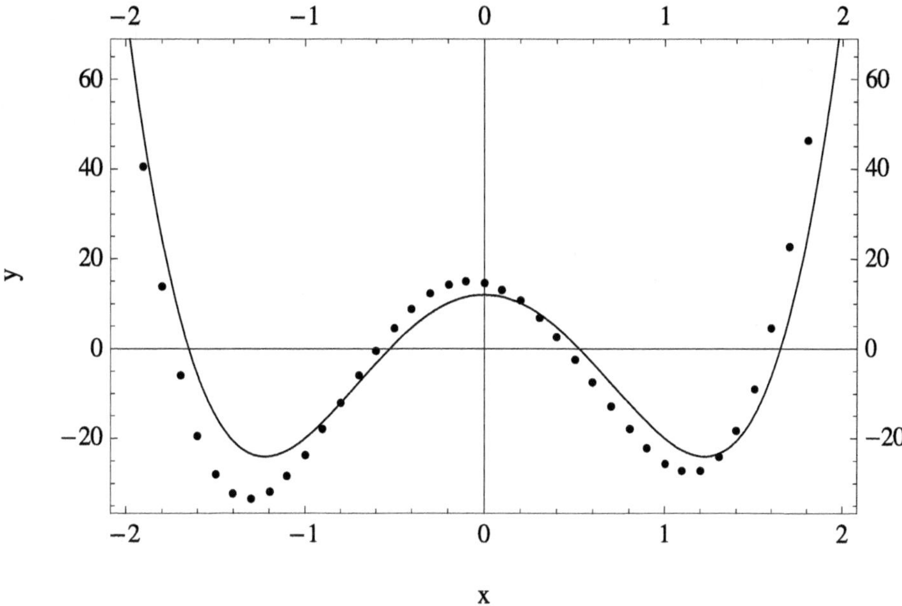

Fig. 2.14 Showing data of Table 2.13 as numerical solution of Hermite differential equation for $n = 4$ for $y_a = +76$ and chosen value of $V = -400$ for $x = -2$. The curve is for known exact result. Using program number 2.2

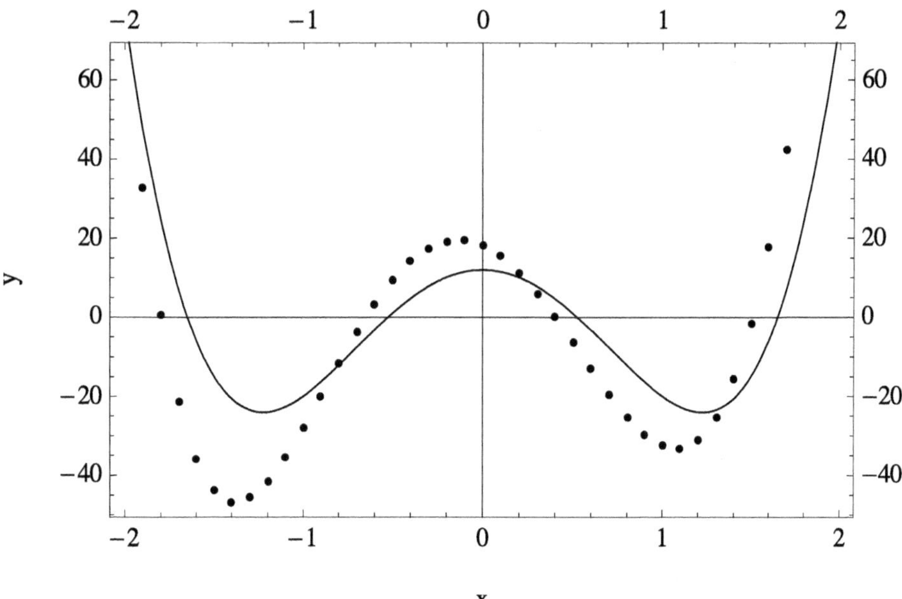

Fig. 2.15 Showing data of Table 2.14 as numerical solution of Hermite differential equation for $n = 4$ for $y_a = +76$ and chosen value of $V = -500$ for $x = -2$. The curve is for known exact result. Using program number 2.2

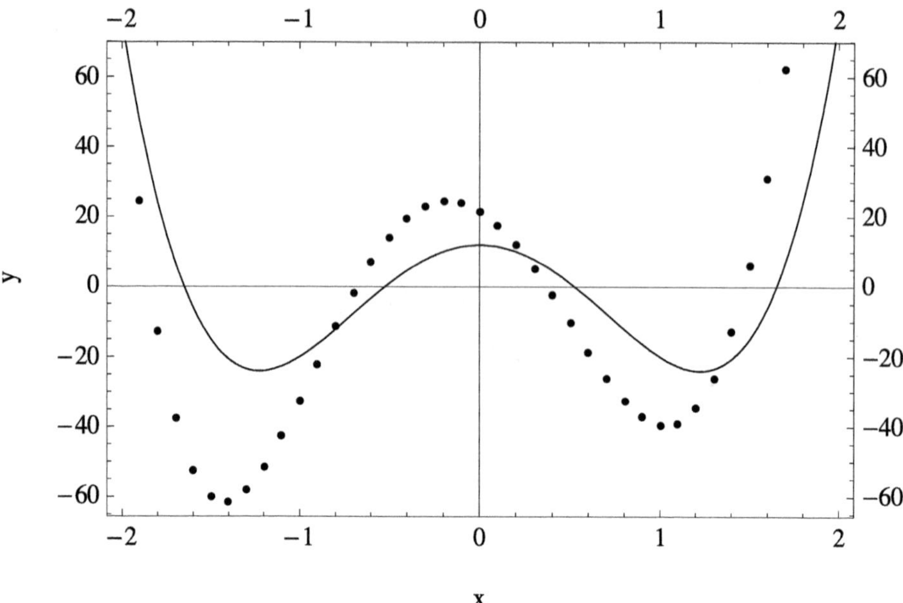

Fig. 2.16 Showing data of Table 2.15 as numerical solution of Hermite differential equation for $n = 4$ for $y_a = +76$ and chosen value of $V = -600$ for $x = -2$. The curve is for known exact result. Using program number 2.2

Table 2.16 Showing values of y called y_r gathered from last rows of Tables 2.9, 2.10, 2.11, 2.12, 2.13, 2.14, and 2.15. We have values of y_r for various chosen values of V at $x = -2$ in the interval 0 to -600

V	y_r	$d = y_b - y_r = 76 - y_r$
0	−63.0969	139.0969
−100	−19.6291	95.6291
−200	23.8387	52.1613
−300	67.3064	8.6936
−400	110.774	−34.774
−500	154.242	−78.242
−600	197.71	−121.71

```
{x,-2,2-h,h}];
TableForm[%,TableSpacing->{2,2},
TableHeadings->{None,{"i","x","V","y"}}]

i=0;
p1=ListPlot[Table[{i=i+1;x=x+h,Y[i]},{x,-2,2-h,h}],
Frame->True,FrameLabel->{"x","y"},FrameTicks->All,PlotStyle->{Black}];

p2=Plot[{12-48*x^2+16*x^4},{x,-2,2},PlotStyle->
{{Black},{Dashed,Black}}];
Show[p1,p2]
```

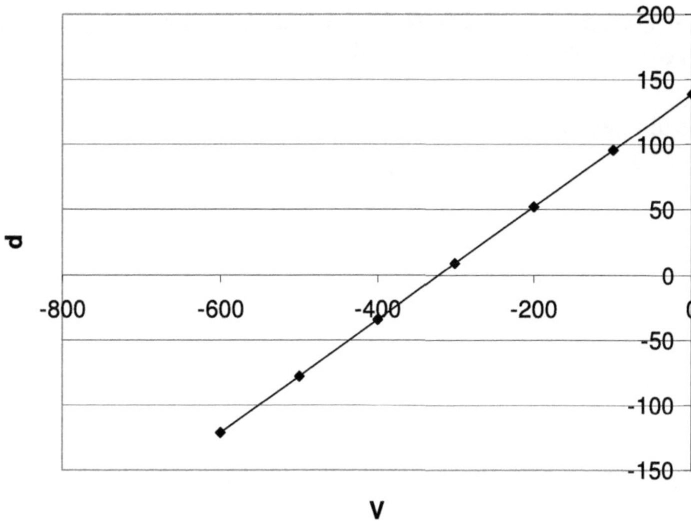

Fig. 2.17   Showing $d = y_b - y_r = 76 - y_r$ of Table 2.16 as a function of various chosen values of V for $x = -2$. We find that for $V = -320$, we have $d = 0$ i.e. $y_b = y_r$

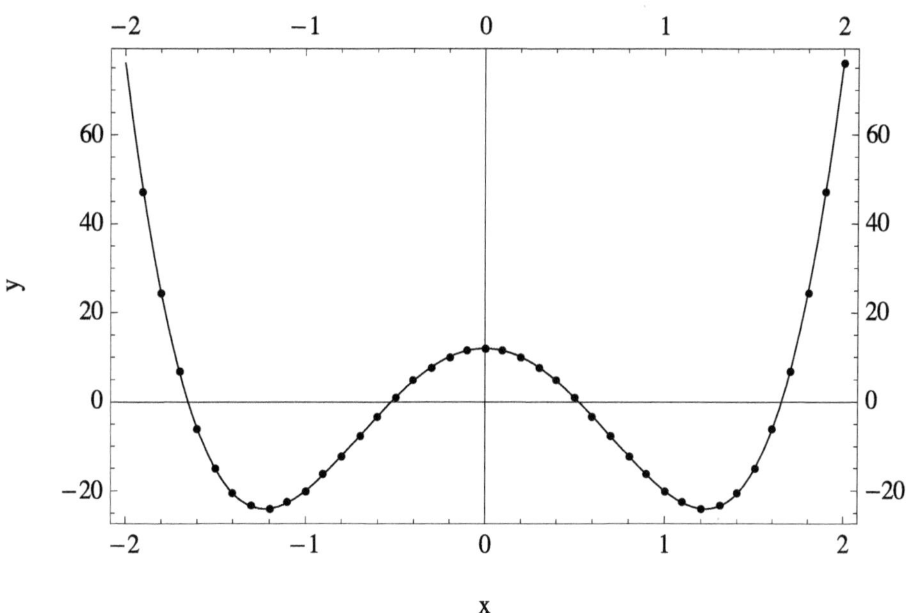

Fig. 2.18   Showing numerical solution of Hermite differential equation for $n = 4$ for given $y_a = +76$ and $V = -320$ for $x = -2$ obtained using Shooting method using program number 2.2. The curve is for known exact result: $y = H_4 = 12 - 48x^2 + 16x^4$; agreement is perfect

2.3 Reproducing Hermite Polynomial H_5 Using Shooting Method Involving Taylor Series Method

The Hermite differential equation is

$$\frac{d^2 y}{dx^2} - 2x\frac{dy}{dx} + 2ny = 0$$

This has well known polynomial solutions for integer values of n. For $n = 5$, the solution is known to be $H_5 = 15x - 20x^3 + 4x^5$. We are given the two boundary values: $y_a = H_5$ $(x = -2) = +2$ and $y_b = H_5 (x = +2) = -2$. We wish to solve the differential equation using Taylor series method using the boundary value y_a as one initial value. The other initial value is the value of $V = \dfrac{dy}{dx}$ for $x = -2$ which we will find out by trial and error. We are assuming that the function $H_5 = 15x - 20x^3 + 4x^5$ is unknown. Correct initial value of V will yield correct value of y for $x = +2$. This is Shooting method involving Taylor series method.

Taylor series is known to be

$$y(x+h) = y(x) + (h/1!)y^I(x) + (h^2/2!)y^{II}(x)$$
$$+ (h^3/3!)y^{III}(x) + (h^4/4!)y^{IV}(x) + (h^5/5!)y^V(x)$$

retaining terms containing upto fifth derivative. As to the Hermite differential equation $\frac{d^2 y}{dx^2} - 2x\frac{dy}{dx} + 2ny = 0$ with $n = 5$, we have

$$V^I = 2xV - 10y \tag{2.6}$$

and hence

$$V^I = 2xV - 10y$$

$$V^{II} = 2xV^I - 8V$$

$$V^{III} = 2xV^{II} - 6V^I$$

$$V^{IV} = 2xV^{III} - 4V^{II}$$

$$V^V = 2xV^{IV} - 2V^{III}$$

where

$$\frac{dy}{dx} = V \tag{2.7}$$

and V^I, V^{II}, V^{III}, V^{IV}, V^V are first, second, third, fourth and fifth derivatives of V with respect to x respectively; we denote these as vd1, vd2, vd3, vd4, vd5 respectively.

Taylor series solution for Eq. (2.6): $V^{\text{I}} = 2xV - 10y$ is given by

$$V_{\text{new}} = V_{\text{old}} + (h/1!)V^{\text{I}} + (h^2/2!)V^{\text{II}} + (h^3/3!)V^{\text{III}} + (h^4/4!)V^{\text{IV}} + (h^5/5!)V^{\text{V}}$$

retaining terms up to fifth derivative. And Taylor series solution for Eq. (2.7): $\dfrac{dy}{dx} = V$ is given by

$$y_{\text{new}} = y_{\text{old}} + (h/1!)y^{\text{I}} + (h^2/2!)y^{\text{II}} + (h^3/3!)y^{\text{III}} + (h^4/4!)y^{\text{IV}} + (h^5/5!)y^{\text{V}}$$

retaining terms up to fifth derivative. Here

$$y^{\text{I}} = V$$

$$y^{\text{II}} = V^{\text{I}}$$

$$y^{\text{III}} = V^{\text{II}}$$

$$y^{\text{IV}} = V^{\text{III}}$$

$$y^{\text{V}} = V^{\text{IV}}$$

which we denote by yd1, yd2, yd3, yd4, yd5 respectively.

We have written program number 2.3 to obtain values of y for $x = +2$ using the given value y_a as the value of y for $x = -2$. We have run the program for various chosen values of V at $x = -2$. Accordingly we have obtained the data tabulated as Tables 2.17, 2.18, 2.19, 2.20, 2.21, 2.22, and 2.23 (See Figs. 2.19, 2.20, 2.21, 2.22, 2.23, 2.24, and 2.25). From the bottom rows of these seven tables, we have gathered values of y for $x = +2$ and we have denoted these as y_r shown in Table 2.24 (See Figs. 2.26 and 2.27).

Program Number 2.3 (for Hermite polynomial H_5)

```
  h=0.1;
x=-2;
y=2;
v=95;
i=0;

Table[{
i=i+1,
x=x+h,

vd1=2*(x-h)*v-10*y;
vd2=2*(x-h)*vd1-8*v;
vd3=2*(x-h)*vd2-6*vd1;
vd4=2*(x-h)*vd3-4*vd2;
```

Table 2.17 Numerical solution of Hermite differential equation for $n = 5$ for $y_a = +2$ and chosen value of $V = 0$ at $x = -2$. Using program number 2.3

i	x	V	y
1	−1.9	−1.6315	1.9125
2	−1.8	−2.6383	1.6945
3	−1.7	−3.1636	1.4009
4	−1.6	−3.3236	1.0739
5	−1.5	−3.2130	0.7452
6	−1.4	−2.9098	0.4377
7	−1.3	−2.4784	0.1675
8	−1.2	−1.9720	−0.0555
9	−1.1	−1.4340	−0.2259
10	−1	−0.8998	−0.3424
11	−0.9	−0.3976	−0.4069
12	−0.8	0.0508	−0.4237
13	−0.7	0.4290	−0.3991
14	−0.6	0.7258	−0.3406
15	−0.5	0.9345	−0.2569
16	−0.4	1.0524	−0.1568
17	−0.3	1.0806	−0.0494
18	−0.2	1.0235	0.0565
19	−0.1	0.8887	0.1527
20	0	0.6863	0.2320
21	0.1	0.4293	0.2882
22	0.2	0.1325	0.3165
23	0.3	−0.1873	0.3139
24	0.4	−0.5114	0.2789
25	0.5	−0.8201	0.2121
26	0.6	−1.0926	0.1161
27	0.7	−1.3073	−0.0045
28	0.8	−1.4425	−0.1427
29	0.9	−1.4764	−0.2896
30	1	−1.3880	−0.4339
31	1.1	−1.1574	−0.5624
32	1.2	−0.7668	−0.6600
33	1.3	−0.2008	−0.7099
34	1.4	0.5517	−0.6940
35	1.5	1.4971	−0.5932
36	1.6	2.6347	−0.3881
37	1.7	3.9545	−0.0601
38	1.8	5.4341	0.4081
39	1.9	7.0345	1.0307
40	2	8.6936	1.8170

Table 2.18 Numerical solution of Hermite differential equation for $n = 5$ for $y_a = +2$ and chosen value of $V = 50$ at $x = -2$. Using program number 2.3

i	x	V	y
1	−1.9	30.3019	5.9775
2	−1.8	14.8303	8.2012
3	−1.7	3.0499	9.0666
4	−1.6	−5.5364	8.9176
5	−1.5	−11.3904	8.0504
6	−1.4	−14.9404	6.7163
7	−1.3	−16.5835	5.1258
8	−1.2	−16.6857	3.4510
9	−1.1	−15.5835	1.8288
10	−1	−13.5841	0.3641
11	−0.9	−10.9662	−0.8675
12	−0.8	−7.9802	−1.8169
13	−0.7	−4.8484	−2.4587
14	−0.6	−1.7657	−2.7882
15	−0.5	1.1006	−2.8190
16	−0.4	3.6101	−2.5800
17	−0.3	5.6498	−2.1127
18	−0.2	7.1332	−1.4686
19	−0.1	8.0009	−0.7066
20	0	8.2198	0.1099
21	0.1	7.7834	0.9155
22	0.2	6.7112	1.6454
23	0.3	5.0492	2.2380
24	0.4	2.8693	2.6379
25	0.5	0.2694	2.7979
26	0.6	−2.6270	2.6819
27	0.7	−5.6708	2.2675
28	0.8	−8.6875	1.5486
29	0.9	−11.4772	0.5375
30	1	−13.8153	−0.7318
31	1.1	−15.4524	−2.2022
32	1.2	−16.1148	−3.7899
33	1.3	−15.5046	−5.3828
34	1.4	−13.3008	−6.8377
35	1.5	−9.1593	−7.9783
36	1.6	−2.7141	−8.5928
37	1.7	6.4216	−8.4315
38	1.8	18.6540	−7.2053
39	1.9	34.4069	−4.5834
40	2	54.1180	−0.1920

Table 2.19 Numerical solution of Hermite differential equation for $n = 5$ for $y_a = +2$ and chosen value of $V = 100$ at $x = -2$. Using program number 2.3

i	x	V	y
1	−1.9	62.2353	10.0425
2	−1.8	32.2989	14.7080
3	−1.7	9.2634	16.7323
4	−1.6	−7.7493	16.7613
5	−1.5	−19.5677	15.3555
6	−1.4	−26.9711	12.9949
7	−1.3	−30.6885	10.0841
8	−1.2	−31.3994	6.9574
9	−1.1	−29.7329	3.8834
10	−1	−26.2684	1.0707
11	−0.9	−21.5349	−1.3280
12	−0.8	−16.0111	−3.2100
13	−0.7	−10.1257	−4.5182
14	−0.6	−4.2572	−5.2358
15	−0.5	1.2666	−5.3812
16	−0.4	6.1678	−5.0033
17	−0.3	10.2189	−4.1761
18	−0.2	13.2430	−2.9938
19	−0.1	15.1132	−1.5660
20	0	15.7534	−0.0122
21	0.1	15.1374	1.5428
22	0.2	13.2899	2.9742
23	0.3	10.2856	4.1621
24	0.4	6.2501	4.9969
25	0.5	1.3590	5.3836
26	0.6	−4.1614	5.2476
27	0.7	−10.0344	4.5395
28	0.8	−15.9325	3.2399
29	0.9	−21.4781	1.3647
30	1	−26.2427	−1.0298
31	1.1	−29.7475	−3.8419
32	1.2	−31.4628	−6.9197
33	1.3	−30.8084	−10.0556
34	1.4	−27.1532	−12.9814
35	1.5	−19.8156	−15.3635
36	1.6	−8.0629	−16.7974
37	1.7	8.8887	−16.8029
38	1.8	31.8740	−14.8186
39	1.9	61.7792	−10.1974
40	2	99.5424	−2.2009

Table 2.20 Numerical solution of Hermite differential equation for $n = 5$ for $y_a = +2$ and chosen value of $V = 150$ at $x = -2$. Using program number 2.3

i	x	V	y
1	−1.9	94.1687	14.1075
2	−1.8	49.7674	21.2147
3	−1.7	15.4768	24.3980
4	−1.6	−9.9621	24.6050
5	−1.5	−27.7451	22.6607
6	−1.4	−39.0017	19.2735
7	−1.3	−44.7935	15.0424
8	−1.2	−46.1131	10.4638
9	−1.1	−43.8824	5.9381
10	−1	−38.9527	1.7772
11	−0.9	−32.1035	−1.7886
12	−0.8	−24.0420	−4.6032
13	−0.7	−15.4031	−6.5778
14	−0.6	−6.7487	−7.6834
15	−0.5	1.4326	−7.9434
16	−0.4	8.7254	−7.4266
17	−0.3	14.7881	−6.2394
18	−0.2	19.3527	−4.5190
19	−0.1	22.2255	−2.4253
20	0	23.2869	−0.1343
21	0.1	22.4915	2.1701
22	0.2	19.8686	4.3030
23	0.3	15.5221	6.0862
24	0.4	9.6308	7.3558
25	0.5	2.4485	7.9693
26	0.6	−5.6959	7.8134
27	0.7	−14.3979	6.8115
28	0.8	−23.1775	4.9311
29	0.9	−31.4789	2.1918
30	1	−38.6701	−1.3277
31	1.1	−44.0426	−5.4817
32	1.2	−46.8108	−10.0495
33	1.3	−46.1121	−14.7284
34	1.4	−41.0057	−19.1251
35	1.5	−30.4720	−22.7487
36	1.6	−13.4117	−25.0021
37	1.7	11.3558	−25.1742
38	1.8	45.0939	−22.4320
39	1.9	89.1516	−15.8115
40	2	144.9670	−4.2098

Table 2.21 Numerical solution of Hermite differential equation for $n = 5$ for $y_a = +2$ and chosen value of $V = 200$ at $x = -2$. Using program number 2.3

i	x	V	y
1	−1.9	126.1020	18.1725
2	−1.8	67.2360	27.7214
3	−1.7	21.6903	32.0637
4	−1.6	−12.1750	32.4487
5	−1.5	−35.9225	29.9658
6	−1.4	−51.0324	25.5521
7	−1.3	−58.8986	20.0007
8	−1.2	−60.8267	13.9702
9	−1.1	−58.0319	7.9927
10	−1	−51.6370	2.4837
11	−0.9	−42.6721	−2.2491
12	−0.8	−32.0730	−5.9964
13	−0.7	−20.6805	−8.6374
14	−0.6	−9.2402	−10.1310
15	−0.5	1.5987	−10.5056
16	−0.4	11.2831	−9.8498
17	−0.3	19.3572	−8.3027
18	−0.2	25.4625	−6.0441
19	−0.1	29.3378	−3.2847
20	0	30.8204	−0.2564
21	0.1	29.8455	2.7974
22	0.2	26.4473	5.6318
23	0.3	20.7585	8.0104
24	0.4	13.0116	9.7148
25	0.5	3.5381	10.5550
26	0.6	−7.2303	10.3791
27	0.7	−18.7614	9.0834
28	0.8	−30.4225	6.6224
29	0.9	−41.4798	3.0189
30	1	−51.0975	−1.6257
31	1.1	−58.3376	−7.1214
32	1.2	−62.1588	−13.1794
33	1.3	−61.4159	−19.4012
34	1.4	−54.8582	−25.2688
35	1.5	−41.1284	−30.1339
36	1.6	−18.7605	−33.2067
37	1.7	13.8229	−33.5456
38	1.8	58.3139	−30.0453
39	1.9	116.5240	−21.4256
40	2	190.3910	−6.2188

Table 2.22 Numerical solution of Hermite differential equation for $n = 5$ for $y_a = +2$ and chosen value of $V = 250$ at $x = -2$. Using program number 2.3

i	x	V	y
1	−1.9	158.0360	22.2375
2	−1.8	84.7045	34.2281
3	−1.7	27.9038	39.7294
4	−1.6	−14.3878	40.2924
5	−1.5	−44.0999	37.2710
6	−1.4	−63.0630	31.8306
7	−1.3	−73.0036	24.9590
8	−1.2	−75.5404	17.4766
9	−1.1	−72.1813	10.0473
10	−1	−64.3213	3.1902
11	−0.9	−53.2408	−2.7097
12	−0.8	−40.1039	−7.3895
13	−0.7	−25.9579	−10.6970
14	−0.6	−11.7316	−12.5786
15	−0.5	1.7647	−13.0678
16	−0.4	13.8408	−12.2731
17	−0.3	23.9264	−10.3661
18	−0.2	31.5722	−7.5693
19	−0.1	36.4501	−4.1440
20	0	38.3539	−0.3785
21	0.1	37.1996	3.4247
22	0.2	33.0260	6.9606
23	0.3	25.9950	9.9345
24	0.4	16.3923	12.0737
25	0.5	4.6276	13.1407
26	0.6	−8.7647	12.9449
27	0.7	−23.1249	11.3554
28	0.8	−37.6675	8.3137
29	0.9	−51.4807	3.8461
30	1	−63.5249	−1.9236
31	1.1	−72.6327	−8.7612
32	1.2	−77.5069	−16.3092
33	1.3	−76.7197	−24.0740
34	1.4	−68.7107	−31.4126
35	1.5	−51.7848	−37.5191
36	1.6	−24.1092	−41.4113
37	1.7	16.2901	−41.9169
38	1.8	71.5338	−37.6587
39	1.9	143.8960	−27.0397
40	2	235.8160	−8.2277

Table 2.23 Numerical solution of Hermite differential equation for $n = 5$ for $y_a = +2$ and chosen value of $V = 300$ at $x = -2$. Using program number 2.3

i	x	V	y
1	−1.9	189.9690	26.3025
2	−1.8	102.1730	40.7349
3	−1.7	34.1173	47.3950
4	−1.6	−16.6007	48.1361
5	−1.5	−52.2773	44.5762
6	−1.4	−75.0937	38.1092
7	−1.3	−87.1087	29.9173
8	−1.2	−90.2541	20.9830
9	−1.1	−86.3308	12.1020
10	−1	−77.0056	3.8967
11	−0.9	−63.8094	−3.1703
12	−0.8	−48.1348	−8.7827
13	−0.7	−31.2352	−12.7565
14	−0.6	−14.2231	−15.0261
15	−0.5	1.9308	−15.6299
16	−0.4	16.3985	−14.6964
17	−0.3	28.4956	−12.4294
18	−0.2	37.6819	−9.0944
19	−0.1	43.5624	−5.0034
20	0	45.8874	−0.5006
21	0.1	44.5536	4.0520
22	0.2	39.6047	8.2894
23	0.3	31.2315	11.8586
24	0.4	19.7730	14.4327
25	0.5	5.7171	15.7264
26	0.6	−10.2991	15.5106
27	0.7	−27.4884	13.6274
28	0.8	−44.9125	10.0050
29	0.9	−61.4815	4.6732
30	1	−75.9523	−2.2216
31	1.1	−86.9277	−10.4009
32	1.2	−92.8549	−19.4390
33	1.3	−92.0234	−28.7468
34	1.4	−82.5632	−37.5563
35	1.5	−62.4412	−44.9042
36	1.6	−29.4580	−49.6160
37	1.7	18.7572	−50.2883
38	1.8	84.7538	−45.2721
39	1.9	171.2690	−32.6538
40	2	281.2400	−10.2367

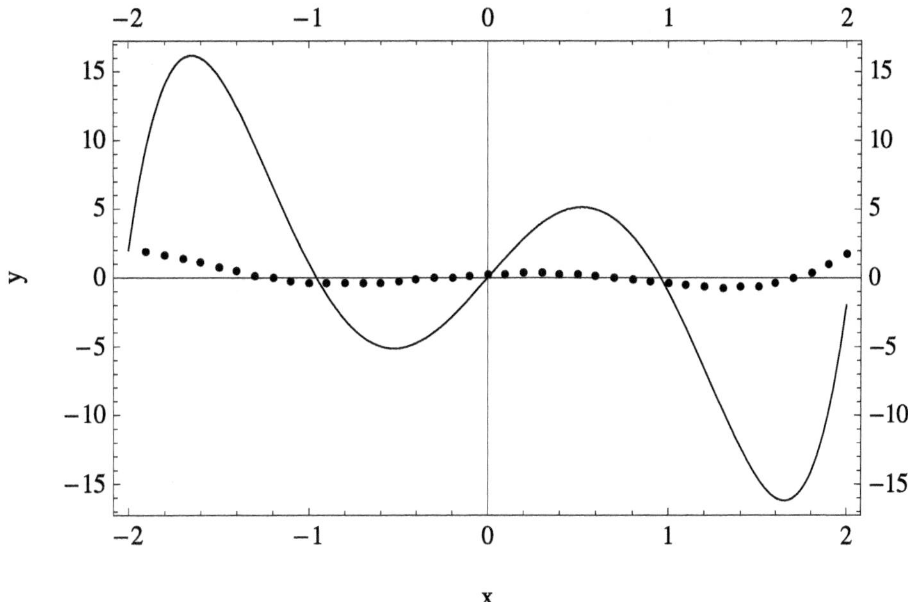

Fig. 2.19 Showing data of Table 2.17 as numerical solution of Hermite differential equation for $n = 5$ for $y_a = +2$ and chosen value of $V = 0$ for $x = -2$. The curve is for known exact result. Using program number 2.3

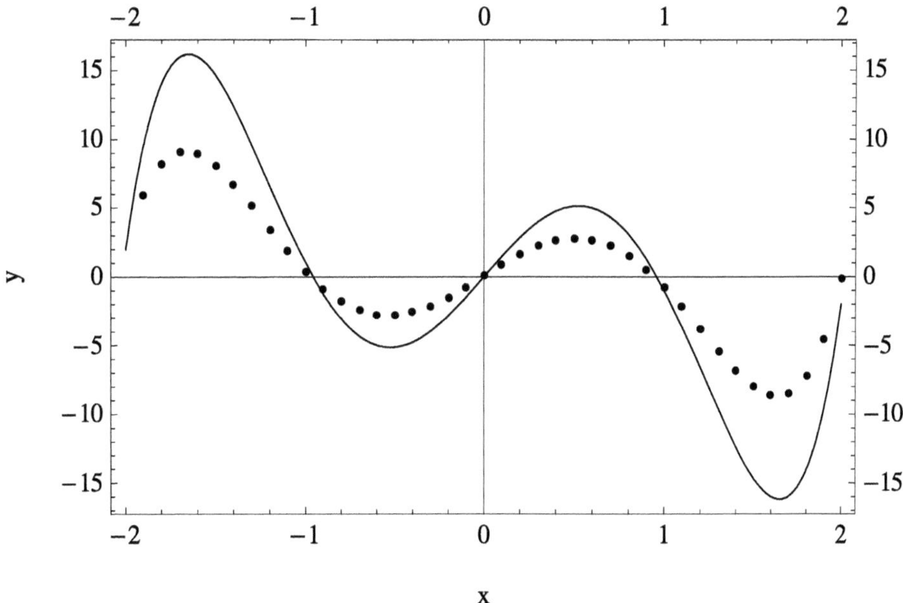

Fig. 2.20 Showing data of Table 2.18 as numerical solution of Hermite differential equation for $n = 5$ for $y_a = +2$ and chosen value of $V = 50$ for $x = -2$. The curve is for known exact result. Using program number 2.3

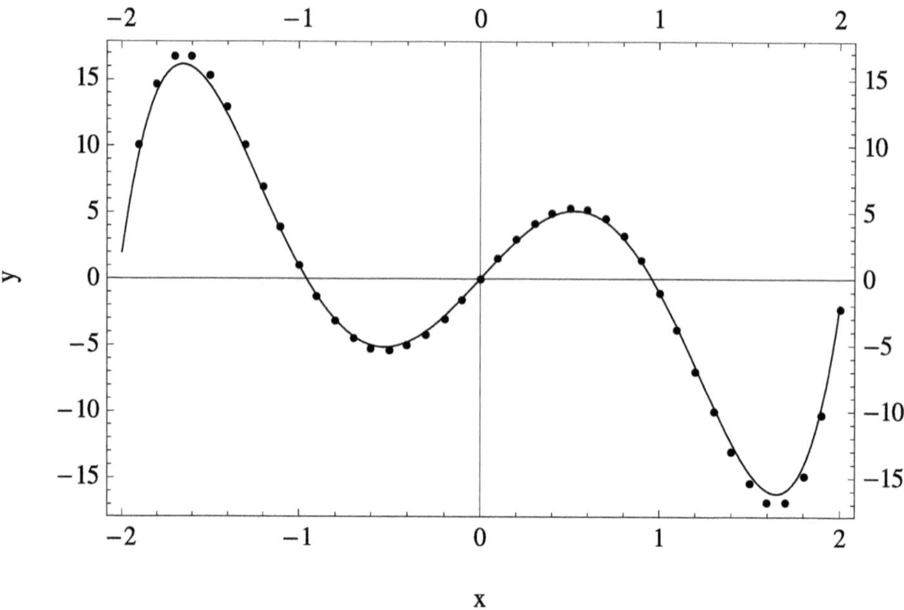

Fig. 2.21 Showing data of Table 2.19 as numerical solution of Hermite differential equation for $n = 5$ for $y_a = +2$ and chosen value of $V = 100$ for $x = -2$. The curve is for known exact result. Using program number 2.3

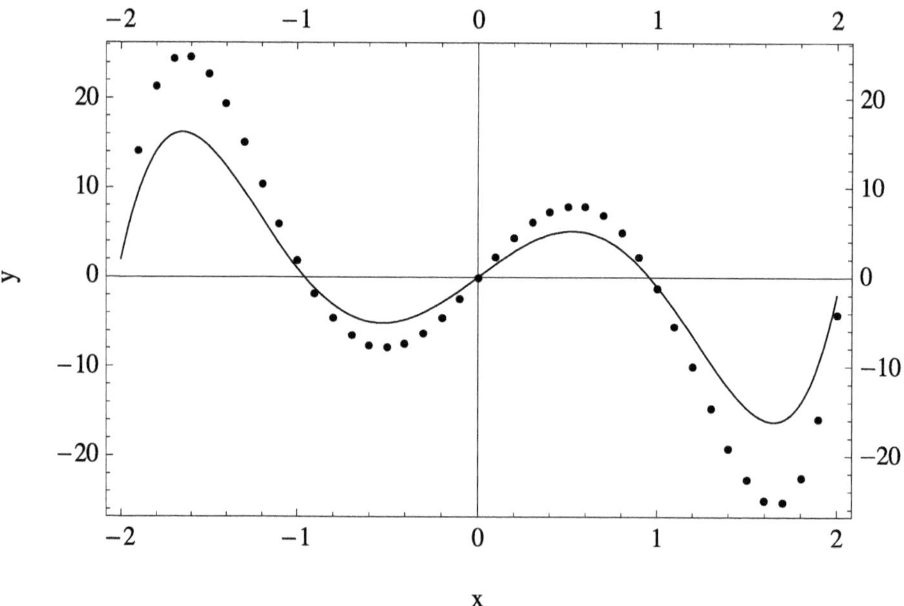

Fig. 2.22 Showing data of Table 2.20 as numerical solution of Hermite differential equation for $n = 5$ for $y_a = +2$ and chosen value of $V = 150$ for $x = -2$. The curve is for known exact result. Using program number 2.3

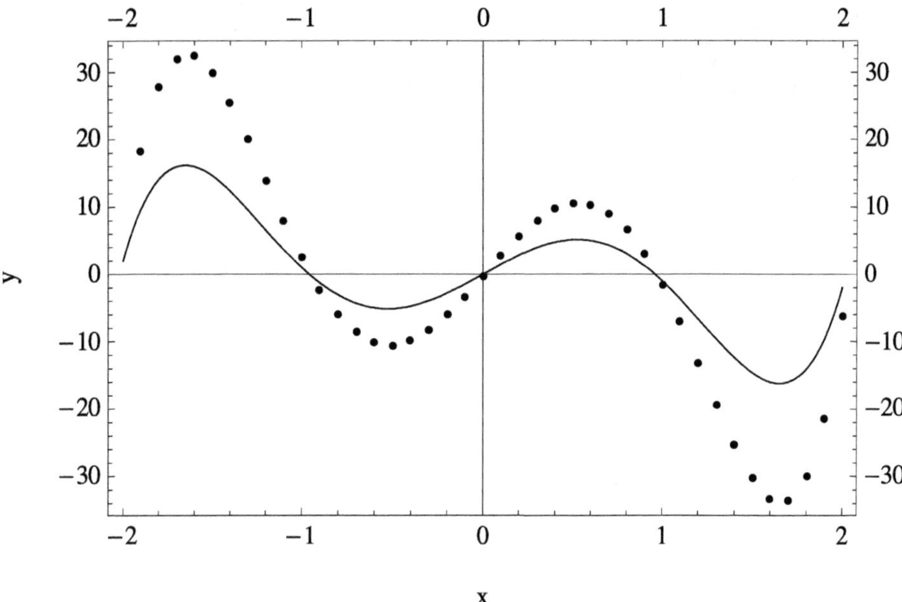

Fig. 2.23 Showing data of Table 2.21 as numerical solution of Hermite differential equation for $n = 5$ for $y_a = +2$ and chosen value of $V = 200$ for $x = -2$. The curve is for known exact result. Using program number 2.3

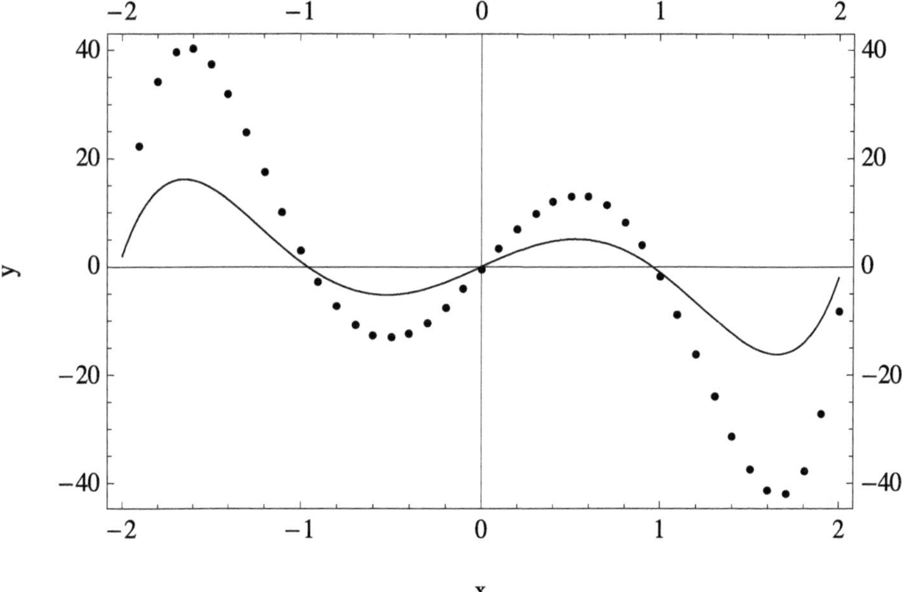

Fig. 2.24 Showing data of Table 2.22 as numerical solution of Hermite differential equation for $n = 5$ for $y_a = +2$ and chosen value of $V = 250$ for $x = -2$. The curve is for known exact result. Using program number 2.3

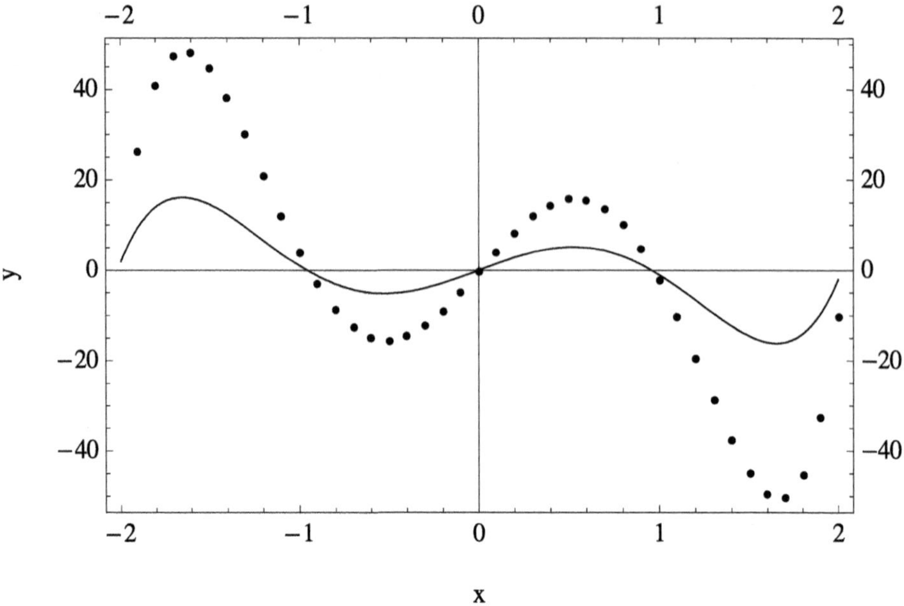

Fig. 2.25 Showing data of Table 2.23 as numerical solution of Hermite differential equation for $n = 5$ for $y_a = +2$ and chosen value of $V = 300$ for $x = -2$. The curve is for known exact result. Using program number 2.3

Table 2.24 Showing values of y called y_r gathered from last rows of Tables 2.17, 2.18, 2.19, 2.20, 2.21, 2.22, and 2.23. We have values of y_r for various chosen values of V at $x = -2$ in the interval 0 to 300

V	y_r	$d = y_b - y_r = -2 - y_r$
0	1.8170	−3.8170
50	−0.1920	−1.8080
100	−2.2009	0.2009
150	−4.2098	2.2098
200	−6.2188	4.2188
250	−8.2277	6.2277
300	−10.2367	8.2367

```
vd5=2*(x-h)*vd4-2*vd3;


yd1=v;
yd2=vd1;
yd3=vd2;
yd4=vd3;
yd5=vd4;


v=v+h*vd1+(1/2)*(h^2)*vd2+(1/6)*(h^3)*vd3+
(1/24)*(h^4)*vd4+(1/120)*(h^5)*vd5,


Y[i]=y=y+h*yd1+(1/2)*(h^2)*yd2+(1/6)*(h^3)*yd3+
```

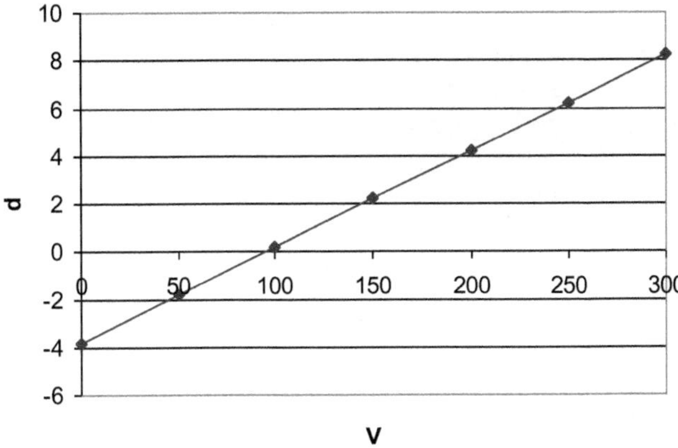

Fig. 2.26 Showing $d = y_b - y_r = -2 - y_r$ of Table 2.24 as a function of various chosen values of V for $x = -2$. We find that for $V = 95$, we have $d = 0$ i.e. $y_b = y_r$

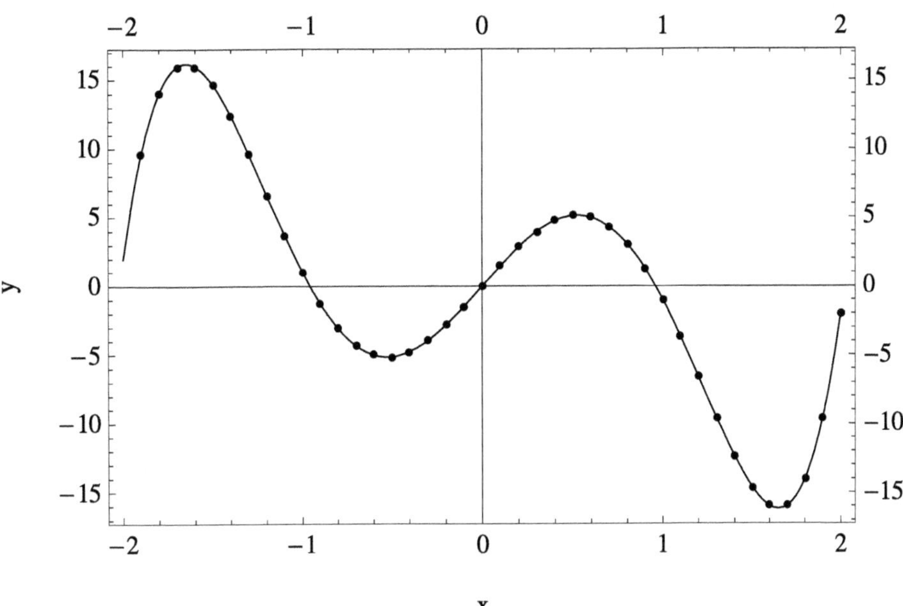

Fig. 2.27 Showing numerical solution of Hermite differential equation for $n = 5$ for given $y_a = +2$ and $V = 95$ for $x = -2$ obtained using Shooting method using program number 2.3. The curve is for known exact result: $y = H_5 = 15x - 20x^3 + 4x^5$; agreement is perfect. Using program number 2.3

```
(1/24)*(h^4)*yd4+(1/120)*(h^5)*yd5},

{x,-2,2-h,h}];
TableForm[%,TableSpacing->{2,2},
TableHeadings->{None,{"i","x","V","y"}}]
```

```
i=0;
p1=ListPlot[Table[{i=i+1;x=x+h,Y[i]},{x,-2,2-h,h}],
Frame->True,FrameLabel->{"x","y"},
FrameTicks->All,PlotStyle->{Black}];


p2=Plot[{15*x-20*x^3+4*x^5},{x,-2,2},
PlotStyle->{Black}];


Show[p1,p2]
```

We have reproduced Hermite polynomials H_3, H_4 and H_5 by numerically solving Hermite differential equation using Shooting method. We have used Taylor series method. We have used different chosen values of $\dfrac{dy}{dx}$ at the left boundary in this regard. We have obtained perfect match between our numerical results and known exact analytic expressions for the polynomials in the interval $-2 < x < +2$. (see Figs. 2.9, 2.18, and 2.27). We have performed symbolic computation using Mathematica® by which we have performed the numerical work.

Concluding Remarks

- We have numerically solved both initial and boundary value problems utilizing Taylor series method.
- There are two chapters: one on initial value problems and another on boundary value problems.
- As to initial value problems, a number of differential equations related to a number of problems of Physics have been solved numerically.
- The equations are first, second and third order differential equations.
- The problems are:
  - radioactive decay,
  - simple harmonic motion,
  - damped harmonic motion,
  - driven damped harmonic motion,
  - motion of oscillators in phase space,
  - cyclotron motion
  - differential equation for Hyperbolic function cosh
- As to oscillatory motion, we have obtained both velocity and displacement of the oscillating particle as functions of time.
- As to third order differential equation, we first turned the second order differential equation obeyed by damped harmonic motion into a third order differential equation. This is intended to explore and check if Taylor series method works well with third order differential equation; and we have found that it does very well.
- As to driven damped oscillations, we found superposition of oscillations of two different frequencies; it fades away with time and steady oscillation of frequency of driving force persists for large values of time.
- As to cyclotron motion, we have numerically solved Newton's second law of motion for Lorentz force exerted on an electron by a uniform static magnetic field, and obtained circular trajectory of electron.

S. Chowdhury, M. G. Moktadir, *Numerical Solutions Using the Taylor Series Method*, Synthesis Lectures on Mathematics & Statistics, https://doi.org/10.1007/978-3-032-26992-8

- We have used Shooting method to solve boundary value problems.
- As to boundary value problems, we have reproduced Hermite polynomials H_3, H_4 and H_5 by numerically solving two-point boundary value problems using so-called Shooting method.
- Use of the Shooting method requires treatment of boundary value problem as an initial value problem, where we have made good use of Taylor series method.
- We have performed symbolic computation in Mathematica® using which we have got the numerical solutions. The programs are evident to course conductors, if not to students.
- A key feature of the results presented in this book is that we can use appreciably large increment of the independent variable and get excellent agreement with known exact analytic solution even for higher (second and third) order differential equations.
- Parts of the book can be used as one of course books on Computational Physics or Computational Mathematics for final year undergraduates of Physics or Mathematics.